INTRODUCTION TO
FLUORESCENCE

INTRODUCTION TO
FLUORESCENCE

David M. Jameson

CRC Press
Taylor & Francis Group
Boca Raton London New York

CRC Press is an imprint of the
Taylor & Francis Group, an **informa** business

CRC Press
Taylor & Francis Group
6000 Broken Sound Parkway NW, Suite 300
Boca Raton, FL 33487-2742

© 2014 by Taylor & Francis Group, LLC
CRC Press is an imprint of Taylor & Francis Group, an Informa business

No claim to original U.S. Government works

Printed on acid-free paper
Version Date: 20150318

International Standard Book Number-13: 978-1-4398-0604-3 (Hardback)

Library of Congress Cataloging-in-Publication Data

Jameson, David M., author.
  Introduction to fluorescence / David M. Jameson.
      pages cm
  Includes bibliographical references and index.
  ISBN 978-1-4398-0604-3 (hardback)
  1. Fluorescence spectroscopy. 2. Biomolecules--Analysis. I. Title.

QP519.9.F56J36 2014
543'.56--dc23   2013041155

Visit the Taylor & Francis Web site at
http://www.taylorandfrancis.com

and the CRC Press Web site at
http://www.crcpress.com

*Dedicated to Gregorio Weber, who taught me that
scientists should put doubt above belief*

Contents

Contents

Preface

ALOHA! FIRST AND FOREMOST, I wish to welcome you to *Introduction to Fluorescence*! I also want to remind you that this book is an *introduction to fluorescence*, not an advanced treatise. The literature already abounds with advanced treatments of each and every topic covered in this book. But having said that, I also want to be clear that this book still requires effort on your part if you wish to acquire a useful understanding of basic fluorescence theory and practice. My former graduate advisor, Gregorio Weber (whom we shall have occasion to mention throughout this book, since his contributions to the field were vast), liked to quote Oscar Wilde, in particular the phrase "The truth is rarely pure and never simple" from the play *The Importance of Being Earnest*. So it is with fluorescence. In an introductory text, I feel obliged to convey the general principles and to use examples that demonstrate these principles. There are exceptions to almost every "rule" proposed in these pages. For example, when I state that fluorescence results from the transition from the first electronic excited state to the ground state, someone may point out that azulene can emit directly from a higher electronic level. Or when I state that the quantum yield of a fluorophore is independent of the exciting wavelength, I may be reminded that tryptophan excited at 230 nm has a lower yield than it does when excited at 280 nm due to photoelectron ejection from the upper electronic level. But in the vast majority of cases, the "rules" I propose hold. If readers of this book demonstrate a sustaining interest in fluorescence and continue to work and study in the field, I have no doubt they will learn of the exceptional cases on their own and may even discover new ones! I should add that clearly not every fluorescence topic or system is covered in this short book. A search of the word "fluorescence" on PubMed, for example, will pull up several hundred thousand articles, which cover hundreds or perhaps thousands of different systems. Hence, it is quite possible that the reader's favorite fluorescence system is not mentioned. Sorry!

For examples of many aspects of fluorescence, I have drawn frequently from my own work, not because my research is particularly interesting (to anyone other than myself!), but rather because I am most familiar with it. I have also mentioned many contributions from Gregorio Weber—in these cases, though, not only am I familiar with his work, but it IS highly significant.

I must acknowledge here my debt to those who helped me, either indirectly or directly, to write this book. It goes without saying that I owe—and will continue to owe—much to Gregorio Weber, from whom I learned so much about fluorescence and also about life. He not only taught in the traditional sense, he also taught by example. He was the great scientific inspiration of my life and, to paraphrase a remark once made by Ludwig (Lenny) Brand (himself a legendary fluorescence practitioner), Gregorio Weber showed by example how scientists should interact with one another, namely with courtesy, generosity, good humor, and humility. During the past four decades, I have also been fortunate to have known and worked with many outstanding scientists from whom I have also learned. Foremost among these is Enrico Gratton, who has been a friend since the day, many years ago, that I picked him up at the airport in Champaign, Illinois, upon his arrival to begin a postdoctoral position in Weber's lab. Enrico is one of the truly outstanding scientists of his generation and his Laboratory for Fluorescence Dynamics, an NIH Research Resource Center, has trained large numbers of students, postdocs, and visiting researchers on advanced fluorescence methodologies. Other friends in Weber's lab during my graduate student days were Bernard Valeur, Antonie Visser, Joseph Lakowicz, Gregory Reinhart, Parkson Chong, and Bill Mantulin, who all went on to make outstanding contributions to the fluorescence field. Many of my own students, postdocs, and collaborators over the years have contributed to my work in the fluorescence field, and I am grateful for their collaborations. I am grateful to my long-term collaborators and friends Joseph Albanesi, Juan Brunet, and Michael Anson for the countless pleasant hours of discussions on myriad topics, including fluorescence. I also want to remember here two former colleagues and friends who passed away while I was working on this book. They are John Eccleston, with whom I collaborated on fluorescence projects for more than 30 years, and Robert Clegg, whom I had known for many years and with whom I shared a passion for history. I miss them both.

I want to thank the present members of my laboratory, postdoc Nicholas James and undergraduate Remy Minami, for reading some of the book in its early form and for offering valuable suggestions. Postdoc Carissa Vetromile, in addition to extensive proofreading, took many of the spectra shown and helped make many of the figures, and I am very grateful for her efforts. I thank Marcin Bury, in my lab, for his excellent and patient photography and proofreading and also for drawing most of the chemical structures in this book. Leonel Malacrida deserves special mention for providing Prodan spectra as well as numerous figures. Greg Reinhart also offered useful suggestions, especially on protein fluorescence. I wish to thank the various funding agencies that have supported my research over the years, including the National Science Foundation, the National Institutes of Health, and the American Heart Association. I am also grateful for the financial support provided in recent years by Allergan, Inc. I also thank the University of Hawaii for providing the intellectual environment that supported this project.

I must acknowledge Taylor & Francis and in particular Luna Han for originally suggesting that I write this book. I also thank Luna for her patience throughout the process—especially as suggested deadlines passed without bearing fruit. I thank Beniamino Barbieri, president of ISS, Inc., who organized many fluorescence workshops over the years and invited me to lecture in them. Organizing and presenting these lectures gave me the opportunity to clarify my thoughts on many topics and the opportunity to discuss material with hundreds of students. I thank the many students who, over the years, endured my lectures on fluorescence. Listening to their questions and comments definitely helped me to shape the contents of this book. I should also acknowledge the Honolulu Bus Company since I spent a great many pleasant hours reading and writing on their buses, which I took from my home in Kailua to the University of Hawaii campus in Manoa valley.

Finally, I wish to acknowledge and sincerely thank Sandra Kopels. She bravely proofread every chapter (and did a really excellent job!) and encouraged and helped me throughout this book project and throughout much of my life. Sandie always inspires me to try harder and makes every day a little nicer for me and everyone around her!

David M. Jameson
Kailua, Hawaii

Author

DAVID M. JAMESON IS PROFESSOR in the Department of Cell and Molecular Biology at the John A. Burns School of Medicine, University of Hawaii at Manoa, having previously served there as professor and chairman of the Department of Biochemistry and Biophysics. He earned a PhD in biochemistry from the University of Illinois at Urbana-Champaign, where his graduate thesis advisor was professor Gregorio Weber. Prior to his move to the University of Hawaii, he was an assistant professor in the Department of Pharmacology at the University of Texas Southwestern Medical Center at Dallas and a post-doctoral fellow at the University of Illinois at Urbana-Champaign and at the CNRS, University of Paris-South in Orsay, France.

Dr. Jameson is the co-organizer of the International Weber Symposia on Innovative Fluorescence Methodologies in Biochemistry and Medicine, which have been held every three years since 1986. He serves on the editorial boards of *Analytical Biochemistry* and *Methods and Applications in Fluorescence,* and he is a member of the advisory board for the Laboratory for Fluorescence Dynamics (a national resource facility supported by the National Institutes of Health) at the University of California, Irvine. He was the recipient of the 2004 Gregorio Weber Award for Excellence in Fluorescence Theory/Application. www.thejamesonlab.wordpress.com.

Introduction

What Is Fluorescence and How Is It Used?

When asked to explain fluorescence to nonscientists, I usually say "When light of a particular color shines on certain types of materials, they give off light of a different color after a very, very short time interval." A slightly more informative definition would be "Fluorescence is the light emitted by an atom or molecule subsequent to the absorption of electromagnetic energy." This criterion, however, is not sufficient to distinguish fluorescence from phosphorescence. So I should add that fluorescence arises from the transition of the excited species from its first excited electronic singlet level to its ground electronic level. Even here there are exceptions. For example, there are rare instances wherein the emitted light comes from higher electronic singlet levels (the classic example being emission from azulene). Also, the phenomenon of "delayed fluorescence" involves a transition from a singlet state to a triplet state followed by a return to the singlet state after a lengthy (in the context of normal fluorescence lifetimes) delay. But throughout this short treatise, I shall strive to discuss the most characteristic phenomena, leaving the exceptions to the specialists.

The fluorescence phenomenon is illustrated in **Figure 1.1**, which shows solutions of several different fluorophores illuminated by a UV handlamp. This figure also illustrates one of the most appealing aspects of fluorescence, namely, that it's pretty! This declaration is not simply a glib description. I remember well when Gregorio Weber—who, as we will learn, made many important contributions to modern fluorescence—explained to me that one of the aspects of fluorescence that initially attracted him was the beauty of the phenomenon and the realization that one could gain a deep appreciation of molecular events simply by observing changes in the hue or intensity of a color accompanying a biochemical process.

Several aspects of fluorescence make it an extremely useful technique for investigations of molecular processes. One of its most salient features is the fact that the excited state persists for a duration in the range of nanoseconds,

FIGURE 1.1 Fluorescent solutions illuminated with a UV handlamp.

that is, some billionths of a second. In essence then, the fluorophore is like a *molecular stopwatch*, which starts with the absorption of light and stops with the emission of light. The fluorophore can thus report on events which transpire on the nanosecond timescale. As we shall learn in the next section, fluorescence has been observed for a long time. In the last couple of decades, however, the popularity of fluorescence techniques has greatly increased. This ever-increasing popularity is largely due to the development of highly sophisticated fluorescent probe chemistries, and the ready commercial availability of these probes, as well as the development of novel microscopy approaches. These considerations, and the emergence of molecular biology methods giving rise to recombinant proteins and site-directed mutagenesis, have led to many novel applications of fluorescence in the chemical, physical, and life sciences. Although the fundamental principles of fluorescence are the same regardless of the application, the examples used in this treatise to illustrate these concepts will be drawn largely from the molecular life sciences including biochemistry and biophysics, clinical chemistry and diagnostics, and cell and molecular biology. The reader will also notice that throughout this discourse I use the first person. This approach is deliberate since I want to direct my writing to the individual reader, as if I were telling her or him a story. Hopefully, this approach will make the learning process more interesting and perhaps more fully engage the reader's attention.

A Nano-History of Fluorescence

A few years ago, I was giving a series of lectures to a group of students who were utilizing fluorescence routinely in research involving microscopy. I asked them when they thought fluorescence was first observed, and after a brief conclave, they announced that the initial observations probably dated to around

1940. It was this misconception that prompted me in future lectures to devote some time to a brief overview of the history of fluorescence.

One of the earliest known discussions of the phenomenon we now recognize as "fluorescence" was due to the Aztecs, as described around 1560 by Bernardino de Sahagún, a Franciscan missionary who actually visited the New World. The Aztecs wrote about a wood which presented some medicinal values and which also conferred unusual colors to water. Specifically, the wood was fashioned into a cup and filled with water. When this water was ingested it conferred some measure of relief to those suffering from urinary problems. The water in the cup also exhibited an intriguing bluish color when viewed in sunlight. Nicolas Monardes, living in Spain, read the reports from the New World and, in 1577, wrote about this medicinal wood and its interesting coloration. The writings of Monardes attracted the attention of others, including an influential Flemish botanist Charles de L'Écluse, who penned a Latin translation in which the wood's name was given as *Lignum nephriticum* (kidney wood), which helped to extend awareness of the strange optical properties throughout Europe. This wood became very popular in the sixteenth and the seventeenth century Europe because of its medicinal virtues for treating kidney ailments. In the ensuing centuries, the wood disappeared from the shops in Europe, and the botanic identity of *L. nephriticum* was lost in a confusion of several species. In 1915, however, W.E. Safford succeeded in disentangling the botanic problem and identified the species which produced the *L. nephriticum* as *Eynsemhardtia polystachia*. In 1982, several highly fluorescent glucosyl-hydroxychalcones were isolated from this plant. Recent studies by A. Ulises Acuña and his colleagues in Madrid demonstrated that the original blue tinge observed by the Aztecs was due to the conversion of Coatline B, under mildly alkaline conditions, to a strongly blue-emitting compound, matlaline (from *matlali*, the Aztec word for blue)—resembling fluorescein—with an emission maximum near 466 nm and with a quantum yield near 1 (which means that virtually every photon absorbed gave rise to a fluorescence photon). The structure of this ancestral fluorophore is shown in **Figure 1.2**. An illustration of its stunning fluorescence is shown in **Figure 1.3**.

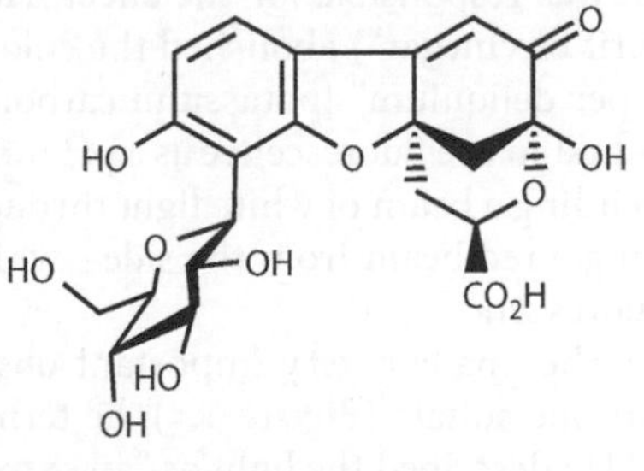

FIGURE 1.2 Structure of matlaline, the fluorescent compound from *Lignum nephriticum*. (Redrawn with permission from A.U. Acuña et al. Structure and formation of the fluorescent compound of *Lignum nephriticum*. *Organic Letters* 11: 3020. Copyright 2009 American Chemical Society)

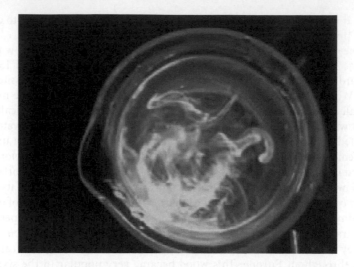

FIGURE 1.3 An alkaline solution containing a piece of *L. nephriticum* wood, illuminated with a UV handlamp.

Many European men of letters were intrigued by the reports on *L. nephriticum* and made significant observations on this system. For example, the polymath German Jesuit priest Athanasius Kircher, among his numerous achievements, wrote a book in 1646 titled *Ars Magna Lucis et Umbrae* (*The Great Art of Light and Shadow*) in which he described his observations on *L. nephriticum*. He noted that light passing through an aqueous infusion of this wood appeared yellow, while light reflected from the solution appeared blue. This observation has prompted some to suggest that Kircher was the "Father of Fluorescence," but, as we shall learn, that title must be reserved for another! Robert Boyle (1670) was inspired by Monardes' report and investigated this system more fully. He discovered that after many infusions, the wood lost its power to give color to the water and concluded that some "essential salt" in the wood was responsible for the effect. He also discovered that addition of acid ("spirit of vinegar") abolished the color and that addition of alkali ("oyl of tartar per deliquium" [potassium carbonate]) brought it back. Hence, Boyle was the first to use fluorescence as a pH indicator! In 1833, David Brewster described sending a beam of white light through an alcohol solution of leaves and observing a red beam from the side—which we now know was due to chlorophyll fluorescence.

In 1845, John Herschel made a very important observation, namely, the fluorescence from quinine sulfate (**Figure 1.4**); he termed this phenomenon "epipolic dispersion." He described the light as "an extremely vivid and celestial blue colour." Intrepid readers can verify this description by looking at tonic water illuminated with a UV handlamp—or, if motivated by a thirst for knowledge, they can visit a drinking establishment with UV illumination handy, and observe the emission from a gin and tonic! (**Figure 1.5**).

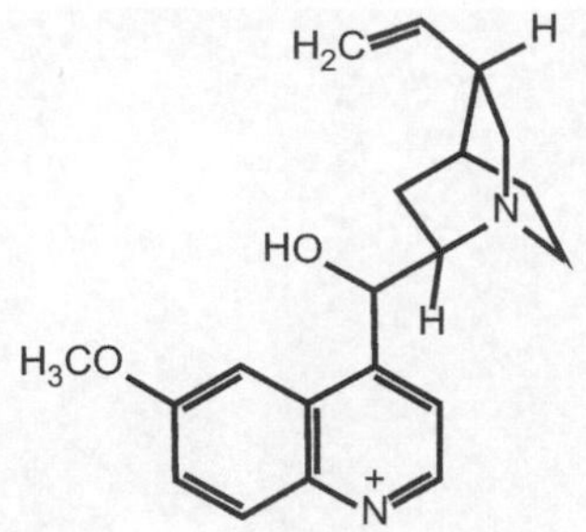

FIGURE 1.4 Structure of quinine.

The title "Father of Fluorescence" must certainly be reserved for George Gabriel Stokes (Figure 1.6), Lucasian Professor at Cambridge University (a position once held by Issac Newton and, until recently, held by Stephen Hawkings) who, in 1852, published his treatise titled *On the Change of Refrangibility of Light*.

In this monumental work, Stokes postulated that the "epipolic dispersion" of quinine sulfate, described by Herschel, was not related to the reflection or refraction of light, but rather was due to the absorption of light followed by the emission of a different color. Stokes used a prism to disperse the solar spectrum and illuminate a solution of quinine and noted that there was no effect until the solution was placed in the ultraviolet region of the spectrum. He wrote: "It was certainly a curious sight to see the tube instantaneously

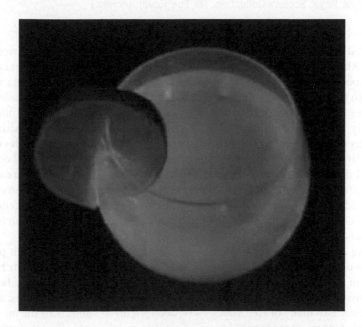

FIGURE 1.5 A gin and tonic illuminated by a UV handlamp.

FIGURE 1.6   Painting of George Gabriel Stokes, 1891, by Hubert von Herkomer, from the Royal Society Collection. (© The Royal Society.)

lighted up when plunged into the invisible rays: it was literally *darkness visible*. Altogether the phenomenon had something of an unearthly appearance." This observation led Stokes to proclaim that the emission was of longer wavelength than the exciting light, which led to this displacement being called the Stokes Shift (although he later acknowledged that Becquerel had independently reached a similar conclusion regarding phosphorescence). Although Stokes originally used the term "dispersive reflection" to describe the emission from quinine sulfate, in a footnote early in his treatise he wrote: "I confess that I do not like this term. I am almost inclined to coin a word, and call the appearance *fluorescence*, from fluor-spar, as the analogous term *opalescence* is derived from the name of a mineral." Thus, we owe the word "fluorescence" to Stokes—without whom the title of this book might have been *Introduction to Dispersive Reflection*.

We should also note that the Englishman William Henry Perkin, played an important, though indirect, role in fluorescence. In 1856, William Henry Perkin was 18 and working as an apprentice in the laboratory of the famous chemist August Wilhelm von Hofman. Interest in quinine was intense at that time since so many of Her Majesty's (Queen Victoria) subjects in the far-flung British Empire were plagued by malaria. Quinine, the only effective treatment at that time, was available from the bark of the cinchona tree, known to grow in Peru and Bolivia (it was introduced as an antimalarial drug in Rome in 1631 by the Jesuit brother Agostino Salumbrino, who had lived in Lima and had observed the Quechua using the bark of the cinchona tree for malarial treatment). Hence, a synthetic path to quinine was highly sought after. Perkin was trying to synthesize quinine from coal tar—not an unreasonable idea given the knowledge of chemistry at that time. In his own words, "I was endeavoring to convert an artificial base into the natural alkaloid quinine, but my experiment, instead of yielding the colourless quinine, gave a reddish powder. With a desire to understand this particular result, a different base of more simple construction was selected, namely, aniline, and in this case obtained a perfectly black powder. This was purified and dried, and when digested with spirits of wine gave the mauve dye" (**Figure 1.7**: note: "mauve" rhymes with "grove"). (*It is interesting to note that it was only in 2001 that Gilbert Stork published the first stereoselective quinine synthesis.*)

Up until this time, virtually all pigments, including fluorescent molecules such as quinine, were derived from natural sources. Chemists had synthesized a few pigments, but Perkin realized the commercial potential of his discovery (especially after Queen Victoria became enamored of the mauve color and wore a dress dyed with mauve to her daughter's wedding) and built the first synthetic dye factory. Soon afterwards, it was recounted that the canal outside his factory was constantly changing color. Perkin is thus rightly considered the father of the synthetic dye industry. One color, however, eluded Perkin and others following in his footsteps and became the "Holy Grail" of the synthetic dye trade; namely, indigo, an important textile dye. In 1878, however, Adolf von Baeyer succeeded in synthesizing indigo (which up until that time was only available as an extract from the plant *Indigofera tinctoria*)—and along the way made a pivotal contribution to fluorescence as discussed below.

Histologists started using the dyes to stain samples within a decade of Perkin's discovery. In 1871, Adolph Von Baeyer synthesized Spiro[isobenzofuran-1(3H), 9′-[9H]xanthen]-3-one, 3′,6′-dihydroxy, which, fortunately

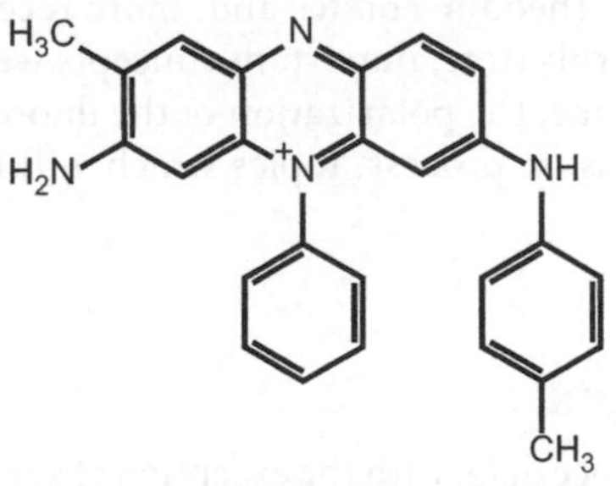

FIGURE 1.7 Structure of the dye mauve (actually mauve is a mixture of mauveine A—the major component, shown here—and mauveine B, which contains an additional methyl group).

for all of us, he named "fluorescein." He apparently coined the name "fluo-rescein," from "fluo" (for fluorescence) and "rescein" (for resorcinol, which he reacted with phthalic anhydride). One of the first scientific applications of fluorescein (which had probably already made the party circuit as a visu-ally arresting demonstration) occurred in 1877 in a major groundwater tracing experiment in southern Germany. Ten kilograms of fluorescein were dumped into the River Danube by A. Kopf and the results showed that the Danube actually flowed to the North Sea (east) rather than into the Black Sea (west) when most of its flow disappeared into its bed near the town of Tuttlingen. In 1961, the tradition began of adding a dye to make the Chicago River green on St. Patrick's Day. For several years fluorescein was used, but now the dye used is apparently a secret!

In 1882, Paul Ehrlich (the famous bacteriologist) carried out the first *in vivo* use of fluorescence when he used uranin (the sodium salt of fluorescein) to track secretion of the aqueous humor in the eye. In 1894, Heinrich Caro, who at that time was working with Ehrlich on new derivatives of methy-lene blue, synthesized the rhodamine class of dyes. In 1887, K. Noack pub-lished a book listing 660 compounds arranged according to the color of their fluorescence (the earliest known example of a Molecular Probes catalog)! R. Meyer, in 1897, used the term "fluorophore" to describe chemical groups which tended to be associated with fluorescence; this word was analogous to "chromophore" which was first used in 1876 by O.N. Witt to describe groups associated with color. Otto Heimstaedt and Heinrich Lehmann (1911–1913) independently developed the first fluorescence microscopes as an outgrowth of the UV microscope. These instruments were soon used to investigate the autofluorescence of bacteria, protozoa, plant and animal tissues, and bio-organic substances such as albumin, elastin, and keratin. In 1914, Stanislav Von Prowazek employed the fluorescence microscope to study dye binding to living cells.

The theoretical foundations of modern, quantitative fluorescence spec-troscopy were established in the first half of the twentieth century by pio-neers, including Otto Stern, Enrique Gaviola, Jean and Francis Perrin (father and son), Peter Pringsheim, Sergei Vavilov, F. Weigert, F. Dushinsky, Alexander Jabłoński, Theodor Förster and, more recently, Gregorio Weber (**Figure 1.8**). During this time, important concepts were introduced such as the excited state lifetime, the polarization of the fluorescence and the quan-tum yield of the emission process, topics which will be discussed in detail in this book.

General Concepts

Virtually all fluorescence data, with the exception of some parameters encoun-tered in fluorescence microscopy, fall into one of the following five categories.

1. The emission spectrum

2. The excitation spectrum

FIGURE 1.8 David Jameson with Gregorio Weber in Hawaii in 1990.

3. The quantum yield

4. The polarization/anisotropy

5. The excited state lifetime

Emission Spectrum

Perhaps, the most common fluorescence parameter measured by novice and expert alike is the emission spectrum. I should first note that it is not unusual to see this parameter referred to as "the fluorescence emission"—a term which seems to me to be somewhat redundant. Unless one must specifically distinguish between the phosphorescence emission and the fluorescence emission, it is usually safe to assume, based on the context, that the "emission spectrum" is referring to the fluorescence process. Basically, the emission spectrum is the record of the variation of the intensity of the emitted light as a function of wavelength, while the sample is illuminated at a fixed excitation wavelength. As we shall learn, we should distinguish between the technical spectrum and the molecular spectrum. The former does not take into account the response characteristics of the detector system, that is, it is "uncorrected," while the latter is "corrected" for the instrument parameters. Later, we shall discuss how emission spectra are "corrected" and when and why one should consider such a correction. In many cases, the emission spectrum allows us to deduce interesting aspects of the system under investigation. For example, the emission maximum may shift in response to a change in the system, such as the fluidity of a membrane, the unfolding of a protein or the chelation of an ion. Sometimes, only the intensity of the emission changes while the emission maximum is invariant. Sometimes, both the intensity and the emission maximum changes, and any change in these parameters may allow for a facile assay for a biological or chemical process, such as ligand/receptor interactions or enzymatic activity.

Excitation Spectrum

To obtain this parameter, the emission wavelength is fixed and the variation of the fluorescence intensity is recorded as the exciting wavelength is changed. Typically, the excitation spectrum encompasses a wide wavelength range, which often, but not always, corresponds to the absorption spectrum of the fluorophore. We shall learn that it is not trivial to obtain a good "corrected" excitation spectrum, that is, one that takes into account the instrument parameters. We shall also learn how excitation spectra can provide clear evidence of Förster resonance energy transfer (FRET).

Quantum Yield

The quantum yield of a fluorophore corresponds to the ratio of the number of photons emitted to the number of photons absorbed. For example, if a solution of a fluorophore absorbs 100 photons and then emits 100 photons its quantum yield is 100% or 1. If the fluorophore emits only 5 of these absorbed photons then its quantum yield is 5% or 0.05. We will learn that most fluorophores of practical utility have quantum yields of at least a few percent and in some cases near 100%.

Polarization/Anisotropy

Polarization and anisotropy are two different, but very related, parameters. Both parameters measure the orientation of the fluorophore with respect to a particular axis, usually the laboratory vertical axis, at the moment that the fluorescence is emitted. These functions report on the movement—or lack thereof—of the fluorophore during its excited state lifetime, that is, during the time between absorption and emission of light. The information content of the polarization and anisotropy functions is essentially identical, and the use of one or the other is usually a matter of convenience and experience. For example, polarization is used almost exclusively in the realm of clinical chemistry whereas anisotropy is more often used in biophysical reports. We shall learn why polarization/anisotropy has become such a popular method in clinical chemistry and drug discovery, and why it is a splendid method to detect protein interactions. As we shall see, polarization/anisotropy is also a useful method to detect FRET.

Excited State Lifetime

The lifetime is nominally a measure of the time between the absorption of a photon by a fluorophore and the subsequent emission of a photon. Just as in radioactive decay, we cannot determine when an individual fluorophore in the excited state will emit a photon, but we can determine with great accuracy the excited state decay parameters of large populations of fluorophores. In fact, the excited state decay of even a single molecule can be measured—but an accurate assessment of the fluorescence lifetime requires that we follow many absorption and emission processes to build up the necessary statistics. We shall learn that a fluorophore's lifetime is an essential parameter which must be taken into account for many types of fluorescence measurements including polarization/anisotropy, quenching, and FRET.

And the Rest

The five categories listed above account for the vast majority of observations. Of course, there are a few specialized techniques, for example, synchronous scanning, and metal enhanced fluorescence, as well as a few more subtle parameters—such as the two-photon cross-section and "molecular brightness"—which have appeared in recent years. In the subsequent chapters, we shall enter into more detail on each topic. Much of this book shall therefore be devoted to describing these five phenomena and discussing how they are accurately determined and what information they provide. Another major topic will be resonance energy transfer, commonly known as Förster resonance energy transfer or FRET. I shall also briefly describe several specialized, yet increasingly popular, techniques, based on fluorescence microscopy including fluorescence recovery after photobleaching (FRAP), fluorescence fluctuation spectroscopy (FFS), fluorescence lifetime imaging (FLIM), super-resolution microscopy and single molecule techniques. It is important to note, however, that anyone planning to utilize these more sophisticated techniques will profit from gaining a solid understanding of the underlying fluorescence principles.

At the end of the book, I include a section on the most common artifacts encountered and the most common errors made by novices. After 40 years "in the trenches," I am reasonably confident that I have seen most fluorescence artifacts and have made most of the common errors—although I also believe that there may still be a few out there lying in wait for me!

Fluorescence methods now permeate the biological sciences. Most of you are reading this book because you want to learn how you can apply fluorescence to your favorite system, be that of an isolated biomolecule or a living cell. I hope I can point you in the right direction or at least give you enough knowledge so that you will know what type of information you can gain using fluorescence and if "the game is worth the candle," that is, if it will be worth your effort! I also hope that I can impart to you some appreciation of the aesthetics of fluorescence. We have all been told that sometimes we should stop to smell the roses—I am suggesting that sometimes we should stop and admire the colors!

Additional Reading

General Fluorescence Texts

B. Valuer and M.N. Berberan-Santos, 2012. *Molecular Fluorescence*, 2nd Edition. Wiley-VCH, Weinheim.

J.R. Lakowicz, 2006. *Principles of Fluorescence Spectroscopy*, 3rd Edition. Springer-Verlag, New York, LLC.

Specific Articles

A.U. Acuña and F. Amat-Guerri, 2008. Early history of solution fluorescence: The *Lignum nephriticum* of Nicolas Monaredes. In *Springer Ser Fluoresc*. Springer-Verlag, Berlin, Vol. 4, pp. 3–20.

A.U. Acuña, F. Amat-Guerri, P. Morcillo, M. Lirasand, and B. Rodríguez, 2009. Structure and formation of the fluorescent compound of *Lignum nephriticum*. *Organic Letters* 11: 3020–3023.

G.G. Stokes, 1852. On the change in refrangibility of light. *Philosophical Transactions of the Royal Society of London*, 142: 463–562.

S. Garfield, 2000. *Mauve: How One Man Invented a Color That Changed the World*. Faber and Faber Limited, London.

D.M. Jameson, 1998. Gregorio Weber, 1916–1997: A fluorescent lifetime. *Biophysical Journal* 75: 419–421.

D.M. Jameson, 2001. The seminal contributions of Gregorio Weber to Modern Fluorescence Spectroscopy. In *New Trends in Fluorescence Spectroscopy*, Eds. B. Valeur and J.-C. Brochon, Springer-Verlag, Berlin, pp. 35–58.

2

Absorption of Light

SINCE FLUOROPHORES REACH their excited state by absorption of incident electromagnetic radiation, it seems reasonable to briefly discuss the nature of light. I should note that other types of excitation modes are possible—such as occurs in bioluminescence, chemiluminescence, or sonoluminescence—but in this book, I shall only consider the case of photoexcitation.

Theories on the nature of light date back to the fifth and the sixth centuries to both India and Greece. The most important modern theories appeared in the seventeenth century and were due to Isaac Newton and Christian Huygens, who championed the corpuscular nature and the wave nature of light, respectively (Rene Descartes and Robert Hooke also published early wave theories of light). We now recognize that light has both wave and particle aspects. In 1865, James Clerk Maxwell published his epochal manuscript, "A dynamical theory of the electromagnetic field," in which he showed that light corresponds to electromagnetic transverse waves whose frequency of vibration (v) and wavelength (λ) are related by $\lambda v = \upsilon$, where υ is the speed of light in the medium of propagation (for a vacuum $\upsilon = c$, where $c = {\sim}3 \times 10^{10}$ cm s^{-1}).

Electromagnetic Radiation: Characterization

Electromagnetic radiation, like any other periodic phenomenon, can be characterized by its amplitude, its frequency, v, and its phase, φ. The frequency is measured in waves per second. The number of waves per centimeter is the wavenumber, $\bar{v}$, related to the frequency, v, by the relation:

$$\bar{v} = \frac{v}{c} \tag{2.1}$$

where c is the speed of light in the medium in which the waves propagate (to be more precise 2.9979×10^{10} cm s^{-1} in a vacuum). The wavelength, λ, is the reciprocal of the wavenumber.

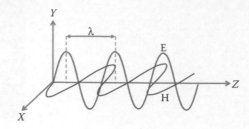

FIGURE 2.1   A depiction of an electromagnetic wave illustrating the electric field (E: blue) and magnetic field (H: red) components propagating in the Z-direction.

$$\lambda = \frac{1}{\overline{\nu}} = \frac{c}{\nu} \tag{2.2}$$

These aspects are illustrated in **Figure 2.1**.

While the frequency is independent of the medium of propagation, the wavenumber and the wavelength are not. The dependence of the velocity of propagation upon the medium gives rise to the phenomenon of dispersion, which is of fundamental importance in spectroscopy.

Photons

According to quantum theory, radiant energy is absorbed or emitted only in discrete amounts or quanta, also called photons in the case of light (the concept of "light quanta" was introduced by Einstein in 1905). The energy, e, of the photon is related to the wave frequency by Planck's relation:

$$e = h\nu = h\frac{c}{\lambda} = hc\overline{\nu} \tag{2.3}$$

where h is Planck's constant, which equals 6.62×10^{-27} erg · s. In photochemical and other applications, it is often convenient to deal with the energy carried by 1 mol of photons, thus connecting the changes of energy occurring in absorption and emission to the energy content per mole of the substance responsible for the absorption or emission. The energy in 1 mol is the Einstein (E):

$$E = Ne = Nh\frac{c}{\lambda} \tag{2.4}$$

with N = Avogadro's number (6.023×10^{23}). If the Einstein is expressed in kg · calories (kcal) and the wavelength of the light in microns, then

$$E = Ne = \frac{28.6}{\lambda}\text{kcal} = \frac{1.24}{\lambda} \tag{2.5}$$

For the center of the visible spectrum $\lambda = 0.5$ μm (500 nm) and $E = 57.2$ kcal or 2.48 eV.

Relevant Wavelength Range

As humans, we are accustomed to viewing a range of wavelengths from about 400 to 700 nm—the so-called visible region (since it is visible to us—but, of course, some species have sensitivities that extend to either side of this range). When I was in elementary school, I learned a simple mnemonic for the colors of the rainbow (**Figure 2.2**), namely "Roy G. Biv" for red, orange, yellow, green, blue, indigo, violet (these seven colors date back to Newton who originally used only five colors for the rainbow but then added orange and indigo to match the seven musical notes in the musical major scale). Interestingly, some of my British friends tell me that they learned "Richard Of York Gave Battle In Vain." Another mnemonic is "Rinse Out Your Granny's Boots In Vinegar," which is, perhaps, less inspiring! In fluorescence studies, we are primarily concerned with the absorption of electromagnetic energy with wavelengths of approximately 200–1000 nm. Quanta in the upper energy limit of this range (~200 nm) carry sufficient energy to produce dissociation of the absorbing molecule into radicals, often followed by irreversible chemical reactions. This circumstance follows from the fact that the chemical bond energies (or bond enthalpies), for single covalent bonds are

FIGURE 2.2   Rainbow over Nawiliwili Bay on Kauai, Hawaii.

typically under 100 kcal mol^{-1}, which corresponds, following Equation 2.5, to $\lambda = 286$ nm. For example, the bond energy of a C–C bond is in the range of 83–85 kcal mol^{-1} whereas a typical C–H bond is close to 100 kcal mol^{-1}. When the shortest wavelengths of this range are used, the instability of the substances under study presents special problems that are not encountered at longer wavelengths, where photochemistry is not such a general phenomenon. The longer wavelength limit of most studies is about 1000 nm due to the rarity of electronic transitions at longer wavelengths (although this restriction does not apply to the special case of multiphoton excitation, which will be discussed in Chapter 9) and by the very strong infrared absorption of water, which restricts observations, at least in the case of biologically important molecules.

Absorption of Light by Molecules

General Considerations

For atoms in the gas phase, the electronic absorption bands are very sharp. The line width is determined solely by the pressure and temperature, and is usually 10^{-2}–10^{-1} Å. The energy levels are sharply separated, and the absorbing states may be completely characterized. For molecules in solution, the situation is considerably more complicated. Instead of an absorption line, a broad band is observed. The broadness originates from the electronic, vibrational, and rotational transitions which may occur. To illustrate the principles involved, let us consider the case of a simple diatomic molecule. The potential energy, E, as a function of the internuclear distance, r, is given by the familiar Morse diagram (**Figure 2.3**).

The minima in the energy level curves correspond to the equilibrium internuclear distance. v_0 is the zero point energy; v_1, v_2... each correspond to the potential energies of a molecule with increasing numbers of quanta of vibrational energy. In the dark, the thermal equilibrium produces a distribution of the molecules among the levels following Boltzmann statistics. The number of molecules in the N_jth level, compared to the number in the ground state, N_0, is given by

$$\frac{N_j}{N_0} = \frac{P_j}{P_0} e^{-\frac{\Delta E}{kT}}$$

(2.6)

where ΔE is the energy difference between the jth and zero level, P_0 and P_j are the degeneracies of the ground and jth states (i.e., the number of states with the same energy), and k is the Boltzmann constant (1.38×10^{-16} erg K^{-1}). Vibrational energy levels are typically separated by some 3 kcal mol^{-1} so that at room temperature only some e^{-6} or about 0.1% are found in the levels above zero. For practical purposes (with some exceptions), and at normal room temperature, we can consider that the absorption of light is always from molecules in the lowest (zero) level of the ground state. Note that the distribution along r is characteristically different for molecules in the zero level than in the other levels. The most probable distance for the zero level is r_0, the equilibrium internuclear distance. For $v > 1$, the distribution becomes bimodal or plurimodal,

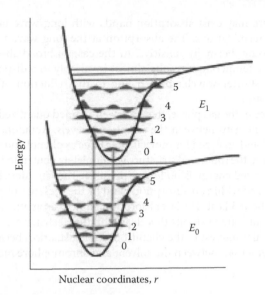

FIGURE 2.3 Morse diagram illustrating two electronic energy levels (E_0 and E_1), and some associated vibrational levels (0–5), as a function of interatomic distance. Absorption (blue) and emission (green) transitions are also depicted.

with the extreme values being the most important. For a certain value, r_d, of r, the interaction becomes negligible and the atoms may pull apart without further change in v. The difference in energy between this state and $v = 0$ is the dissociation energy D of the molecule, which is determined by the bond energy between the atoms.

Franck–Condon Principle

Absorption, or emission, in polyatomic molecules is governed by the Franck–Condon principle, enunciated by James Franck and Edward Condon in 1926. In its simplest form, the Franck–Condon principle may be formulated as follows: Changes in electronic distribution occur very fast in comparison with changes in bond angles and bond distances. Therefore, the nuclear configuration of a molecule, that is, the set of bond distances and bond angles, cannot change appreciably during the process of absorption or emission of light by the molecule. In the simplest case of a diatomic molecule, the Franck–Condon principle means that an electronic transition may be represented by a vertical line starting from r_0 and ending in an excited state with this very same r value, as depicted in **Figure 2.3**. Since $r_0^* > r_0$, it follows that an excited vibrational level is virtually always reached in absorption. Evidently, as depicted in **Figure 2.3**, many upper vibrational levels can be reached provided that the field contains the necessary energy. Some molecules in the ground state have values of r much larger or much smaller than r_0, since there is a distribution of r values in the molecules in the zero level of the ground state.

Hence, there may exist absorption bands with long-wave tails rather than abrupt cutoff terminations. The absorption at the long-wave tails of bands is often found to be thermally sensitive. In the case of broad absorption bands, lowering of the temperature results almost invariably in a displacement of the band edge to shorter wavelengths, owing to the reduction of the thermally excited population.

In molecules in the gas phase, there are well-defined quantized rotational levels, which fill the gaps between the vibrational levels. Rotational structure has been observed and analyzed for molecules as large as phenol, indole, and anthracene. In solution, the rotational structure is completely obliterated since free rotation with quantized energy is no longer possible. The vibrational structure may still be recognized within an absorption band by the presence of peaks separated in energy by about 3 kcal. The last absorption band of aromatic hydrocarbons in nonpolar solvents often exhibits this clear vibrational structure. A further influence, of particular interest to the chemist or biologist, arises because of possible molecular interactions between the solvent and chromophore molecules.

Effect of Conjugation on Absorption

A useful, albeit somewhat simplified, way to consider light absorption by molecules containing conjugation involves molecular orbital theory, which uses a linear combination of atomic orbitals (LCAO) to represent molecular orbitals encompassing the entire molecule. These molecular orbitals may be classified as bonding, antibonding, and nonbonding, as illustrated in **Figure** 2.4, which also depicts electronic transitions between various orbitals.

In the majority of the systems of interest to the fluorescence practitioner, which are usually aromatic molecules, $\pi-\pi^*$ transitions are responsible for light absorption in the lower energy absorption regions (in a few cases $n-\pi^*$ transitions come into play). Since π-orbital bonding occur in carbon–carbon double bonds, it follows that molecules with double bonds are of interest to us, especially when a system of alternating single and double bonds exist, which allows delocalization of π electrons, or *conjugation*. As the conjugation in a molecule increases, the energy of the $\pi-\pi^*$ transition decreases—which corresponds to a red shift (also known as a bathochromic shift) in the absorption, as illustrated in **Figure** 2.5. One notes that the strength of the absorption, or

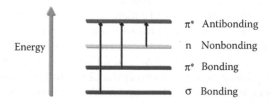

FIGURE 2.4 A depiction of the energies associated with transitions between various molecular orbitals.

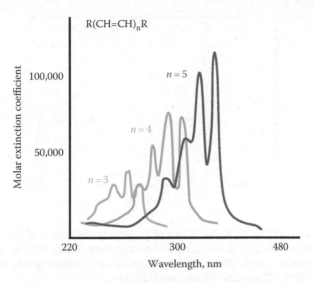

R(CH=CH)$_n$R

FIGURE 2.5 A depiction of the effect of increasing the number of conjugated double bonds on the absorption spectra for several polyenes. (Modified from http://www2.chemistry.msu.edu/faculty/reusch/VirtTxtJml/Spectrpy/UV-Vis/spectrum.htm.)

the extinction coefficient (discussed below) also increases with increasing conjugation. In cases of *cis* and *trans* isomers, the *cis* isomer usually absorbs at shorter wavelengths with a lower extinction coefficient compared to the *trans* isomer.

Such conjugation also occurs in aromatic systems, and again the greater the extent of conjugation the more red-shifted the absorption, and the greater the molar extinction coefficient. **Figure 2.6** depicts these effects for naphthalene, anthracene, and naphthacene. In cases wherein the conjugated system, either linear or aromatic, contains additional chemical groups, for example, carbonyl, amido, or azo groups to name but a few, the absorption characteristics may be altered by two effects. First, the additional chemical group may be chromophoric, hence adding to the absorption at particular wavelengths. For example, a carbonyl group (RHC=O) will exhibit an absorption maximum around 290 nm with a molar extinction coefficient near 16, due to an $n–\pi^*$ transition. These transitions may appear as shoulders on the main absorption band. Second, the chemical group may affect the underlying absorption characteristic of the parent molecule. The effect of these groups will depend on whether they are electron donating or electron withdrawing. As one can imagine, the literature on this topic is considerable and readers with a sustaining interest are referred to classic texts such as Silverstein, Bassler, and Morrill, *Spectrometric Identification of Organic Compounds*. These general considerations are useful when considering a molecular structure and estimating the wavelength region in which it is likely to absorb light—if at all.

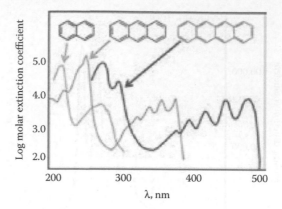

FIGURE 2.6 A depiction of the effect of increasing the length of the conjugated ring system on the absorption spectra of naphthalene (blue), anthracene (green), and naphthacene (red). (Modified from http://www2.chemistry.msu.edu/faculty/reusch/VirtTxtJml/Spectrpy/UV-Vis/spectrum.htm.)

Effects of Molecular Environment on Absorption

Although our primary interest may be the effect of molecular environments on fluorescence parameters, we should realize that environmental effects on absorption also occur, albeit usually to a smaller extent than with fluorescence. Alterations in the solvent polarity may certainly affect the absorption spectral maximum as well as the strength of the absorption (the extinction coefficient), but the changes in these parameters are usually very small (e.g., 5–10 nm in the absorption maximum and <10% in the extinction coefficient) compared to the *potential* changes on the fluorescence maximum and quantum yield (the sensitivity of fluorescence probes to the environment changes dramatically depending on the fluorophores—a topic to be discussed in more detail in Chapter 10). These changes, however, refer to properties of the fluorophores in isotropic environments, that is, when similar molecules surround the probe as in a pure solvent. When fluorophores are embedded in a nonisotropic environment, such as a protein matrix or a membrane, the effects on the absorption and fluorescence properties are not easy to predict. As will be discussed in more detail later in this book, alterations in the environment of the chromophore in fluorescent proteins, such as GFP, can dramatically alter both the absorption and fluorescence properties of the chromophore. It is interesting to note that nature has experimented with protein sequences to alter the absorption properties of protein-associated pigments to produce color vision in mammals. Many of the phenotypes of color blindness originate in the alteration of one or more of the amino acids in the rod cell retinal binding proteins. One can imagine that "tweeking" with the absorption properties of these proteins could have provided advantages to early primates as regards localization/identification of redder (and hence younger and healthier) leaves and ripe fruits, or spotting predators, and so on. Interestingly, since the genes that produce photopigments are carried on the X chromosome, color blindness occurs in a much higher percentage of males compared to females: ~7% of males compared to <1% of females. One may speculate if this fact could account for some of the arguments between men and women regarding color coordination of clothing!

Practical Considerations

Beer–Lambert Law

The ability of a molecule to absorb light at particular wavelengths can be considered in the context of the Beer–Lambert law, sometimes known as the Beer–Lambert–Bouguer law to recognize the contributions of August Beer (a German physicist), Johann Heinrich Lambert (a Swiss astronomer), and Pierre Bouguer (a French astronomer). Interestingly, Pierre Bouguer calculated the attenuation of light passing through the atmosphere in 1729, which anticipated the work by Lambert in 1760 and by Beer in 1854. The Beer–Lambert law may be stated as

$$A = OD = \log_{10}\left(\frac{I_0}{I}\right) = \varepsilon c l \tag{2.7}$$

where A is the absorbance of light through a material of pathlength l and concentration (of the absorbing substance) c. The term ε is the molar absorptivity or, more commonly, the extinction coefficient of the absorbing material. The terms I_0 and I correspond to the intensity of the light entering and exiting the material, respectively. The term OD corresponds to the *optical density*.

Since the log of intensity ratio is unitless, it follows that the units of ε will be M^{-1} cm^{-1}. As a rule of thumb, the useful range over which the absorption of solutions can be accurately determined is generally about 0.01–2.0 optical density (OD) units, depending on the precise instrumentation utilized. The upper limit is usually due to parasitic light in normal spectrophotometers, which results in deviations from linearity as the OD exceeds about 2 (although some instruments have a higher linear range), while the lower limit is generally due to the inherent noise in the instrument. I should note that the most accurate absorbance measurements are usually obtained near an OD of 1. Molecules which strongly absorb light are those with large extinction coefficients, and the highest such values, for simple organic fluorophores, found in typical systems rarely exceed 200,000 M^{-1} cm^{-1} (I note that extinction coefficients near 250,000 $M^{-1}cm^{-1}$ have been reported for Cy5, a fluorophore discussed in Chapter 10). Hence, we can see that with a 1 cm pathlength, the practical concentration range for quantification of biomolecules by absorption is around 10^{-4}–10^{-7} M. We may note that the "transmittance" is the fraction of light absorbed by the sample and is given by: $T = I/I_0$. Hence, $A = -\log_{10}T$. If a sample has an optical density of 1, then 90% of the incident light is absorbed by a 1 cm pathlength solution. Similarly, optical densities of 2 and 0.3 correspond to 99% and 50% light absorption (these considerations will be important when the "inner filter effect" is discussed in Chapter 3) (**Figure 2.7**).

Absorption measurements, such as spectra or optical densities, are determined using a spectrophotometer. Often, though not always, these instruments

FIGURE 2.7 Illustration of optical densities in four solutions of rhodamine B illuminated with a 528 nm penlight (illumination direction from the right side of cuvette). From left to right, the optical densities at the illumination wavelength are 3.0, 1.0, 0.3, and 0.03.

are of the so-called "double-beam" or "dual beam" design, which means that the light absorbed by one solution is compared, in real time, to that absorbed by a "blank" solution. Typically, one places the appropriate solvent blank, for example, buffer, in the reference pathway of the spectrophotometer and the sample in the sample pathway. The instrument then compares the amount of light reaching the photodetector from each pathway and calculates the resulting absorbance (or transmittance) at a specific wavelength or as the wavelength is scanned (**Figure 2.8**). The light sources in such instruments are usually tungsten or halogen lamps for the visible wavelength regions and deuterium lamps for the ultraviolet (below about 340 nm).

If one knows the extinction coefficient of a compound, then the optical density can be used to obtain that compound's concentration. For example, if the optical density, at 490 nm, of an alkaline fluorescein solution is 1.0 (in a normal 1 cm pathlength cuvette), using Equation 2.7 and an extinction coefficient of 80,000 $M^{-1}cm^{-1}$, one calculates a concentration of 1.25×10^{-6} M. OD measurements at 280 nm are excellent for estimating a protein concentration. If one knows the amino acid sequence of the protein, and hence the number of tryptophan, tyrosine, and cysteine residues, then a simple formula will give a good estimation of the protein concentration, based on the measured OD. Specifically, the protein extinction coefficient, worked out empirically (see Pace et al., listed in Additional Reading) is approximately given by

$$E_{280\,nm}\ (M^{-1}cm^{-1}) = (\text{number of Trypt})(5500) + (\text{number of Tyr})\ (1490)$$
$$+ (\text{number of Cys})(125) \tag{2.8}$$

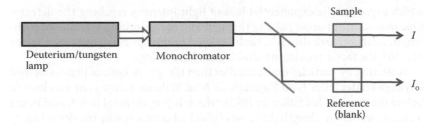

FIGURE 2.8 Schematic diagram of a dual beam UV-Vis spectrophotometer.

Note that the cystine S–S bond contributes slightly to the OD at 280 nm. In measurements on macromolecules, however, one must be wary of light scattering as discussed below.

Departures from Beer–Lambert Law

In practice, one sometimes finds that the linear dependence on concentration predicted by the Beer–Lambert law does not hold. In the early days of commercial spectrophotometers, these deviations were sometimes due to instrument artifacts such as fluorescence or parasitic light (i.e., light at an unwanted wavelength) reaching the detector. Both of these scenarios would result in an artificially low apparent absorbance. For example, if the actual optical density of a solution is 2, then only 1% of the light at that wavelength is reaching the detector. If there is even a fraction of a percent of parasitic light (light at a different wavelength) in the original beam, which is not absorbed, then this spurious light can make a significant contribution to the total light reaching the detector and the apparent optical density will be reduced. This effect makes it problematic to accurately measure optical densities higher than 2 or 2.5, depending on the instrument design. Modern instruments are usually properly designed to minimize such problems—at least we hope they are! If one absolutely must measure the optical density of a strongly absorbing solution, and for some reason cannot dilute the sample, then one can always utilize a cuvette with a short pathlength—these types of cuvettes are shown in Chapter 3—which will reduce the optical density to a manageable range. Also, some new types of absorption instrumentation are now available, which use specialized optics to measure the absorption in an extremely small volume (a few microliters), and hence very short pathlengths.

Scattering of the incident light, that is, turbidity, is a problem often encountered with certain biological samples such as membrane or aggregating protein solutions. One often sees "turbidity" measurements in the literature, for example, to follow protein aggregation processes, and these data are usually presented with "optical density" or "absorbance" plotted on the Y-axis. Turbidity is, in fact, closely related to these terms, but the definition of turbidity (Tb) is given by

$$I = I_0 e^{-l\text{Tb}} \tag{2.9}$$

which expresses the exponential loss of light intensity reaching the detector (due to scattering) upon passing through any nonabsorbing medium of pathlength *l*. Hence, turbidity is actually related to the optical density by a factor of 2.303, the factor relating natural logs to base 10 logs.

Scattering by particles much smaller than the wavelength of light is termed Rayleigh scatter (after Lord Rayleigh, or John William Strutt as he was known before the death of his father in 1873), which is proportional to λ^{-4}, and hence increases as the wavelength decreases (which of course is why the sky is blue— but that's another story)! Rayleigh scatter thus characteristically increases as the wavelength decreases in an absorption scan—as indicated in **Figure 2.9** with a dilute solution of glycogen. When scanning an aggregated protein solution, the protein absorption essentially sits on top of the scatter curve (**Figure 2.9**), and scattering must be taken into account if one intends to determine the protein concentration from the absorption measurement. To a good approximation, if one can accurately measure the solution's absorption at 330 nm, one can multiply that value by 1.926, and subtract the product from the absorption at 280 nm to correct for light scattering (see Pace et al. in Additional Reading). In the case of larger scattering particles, one enters the realm of Mie scattering, which we shall not consider in this introductory treatise.

Another affront to the honor of Messieurs Beer and Lambert can arise from the formation of molecular complexes. Already in 1889, B. Walter discovered that the fluorescence from fluorescein solutions did not increase indefinitely with the concentration, and speculated that some sort of aggregation process was responsible. In 1909, S.E. Sheppard wrote a seminal paper demonstrating that concentrated solutions of isocyanine dyes showed a marked deviation from the Beer–Lambert law, and specifically called attention to the appearance

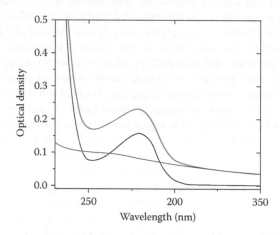

FIGURE 2.9 Effect of turbidity on the absorption spectrum of a protein solution. The black spectrum corresponds to a solution of bovine serum albumin (BSA) while the blue spectrum is for the same BSA solution with glycogen added to produce scattering. The red spectrum corresponds to the same amount of glycogen in the absence of protein. (The author would like to thank Carissa Vetromile for these spectra.)

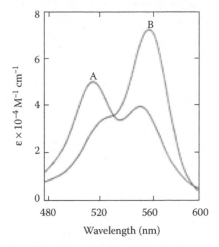

FIGURE 2.10 Absorption spectra for tetramethylrhodamine solutions illustrating monomer and dimer (excimer) absorption. Spectrum A corresponds to tetramethylrhodamine dimers while spectrum B corresponds to tetramethylrhodamine monomers. (Originally published in the *J. Biol. Chem.* B.D. Hamman et al., 1996. Tetramethylrhodamine dimer formation as a spectroscopic probe of the conformation of *Escherichia coli* ribosomal protein L7/L12 dimers. 271: 7568–7573. © the American Society for Biochemistry and Molecular Biology.)

of the absorption spectra of isocyanine in organic and water solutions. He noted that isocyanine, in alcohol, exhibited a maximum at 575 nm and a small shoulder at 535 nm, while in water the 535 nm band became a separate band, comparable in intensity to the 575 nm band. Thus began the formal study of dye aggregation. Aggregation of dyes, most notably xanthene-based dyes such as rhodamines, have been useful in certain types of modern fluorescence studies and will be discussed later in this book. For now, though, I shall simply show, in **Figure 2.10**, two absorption spectra due to rhodamine—in one case (a), they form ground-state dimers, termed excimers, while in the other case (b), the rhodamine molecules do not interact with each other. These types of excimers are often not fluorescent, which can be a help or a hindrance, depending on one's system and experimental goals.

Additional Reading

R.M. Silverstein, G.C. Bassler, and T.C. Morrill, 1991. *Spectrometric Identification of Organic Compounds.* John Wiley & Sons, Inc., New York.

C.N. Pace, F. Vajdos, L. Fee, G. Grimsley, and T. Gray, 1995. How to measure and predict the molar absorption coefficient of a protein. *Protein Sci.* 4: 2411–2423.

B.D. Hamman, A.V. Oleinikov, G.G. Jokhadze, D.E. Bochkariov, R.R. Traut, and D.M. Jameson, 1996. Tetramethylrhodamine dimer formation as a spectroscopic probe of the conformation of *Escherichia coli* ribosomal protein L7/L12 dimers. *J. Biol. Chem.* 271: 7568–7573.

Instrumentation

As discussed in Chapter 1, formal fluorescence measurements have been made for over 150 years. Fluorescence instrumentation, in the more modern sense, has been around for almost a century and has been commercially available for more than half a century. The earliest commercial instruments were attachments for spectrophotometers, such as the Beckman DU spectrophotometer; this attachment allowed the emitted light (excited by the mercury vapor source through a filter) to be reflected into the spectrophotometer's monochromator. The first commercial spectrofluorimeters with monochromators for both excitation and emission, and with quartz optics, were inspired by the work of Robert Bowman at the National Institutes of Health in the early 1950s, and were produced by Aminco (American Instrument Company) in 1956. The Farrand Optical Company also produced a commercial spectrofluorimeter about this time. These early spectrofluorimeters allowed biologists to use fluorescence to develop clinically relevant assays for a wide variety of biological molecules.

All fluorescence observations require certain components, namely,

1. A light source

2. Light selection devices

3. A light detector

The components required for quantitative fluorescence observations are sketched below.

The precise nature of these components changes over time as new technologies appear. Early observations on fluorescence were carried out using the sun as the light source and the eye as the detector. In fact, visual observations were the principal means of detection and quantification of fluorescence until well into the twentieth century. Reliable and sensitive photomultiplier tubes, the primary light detectors used today, only came into general use after World War II. Gregorio Weber, the father of modern fluorescence, once told me that early photodetectors were so electronically noisy that some people doubted that they would replace the human eye in fluorescence determinations! The danger that direct observation of infrared and ultraviolet radiation posed to human vision was not fully appreciated by the pioneers and consequently many of them, including Gregorio Weber, suffered from acute eye ailments, such as detached retinas and cataracts, in later life. Hence, a warning to fluorescence practitioners—*avoid direct observation of UV and IR light*! (This advice is in addition to the well-known warning posted in many laboratories using lasers, namely: "Do not look into laser with remaining eye!"). Advances in optics, light sources, and detectors during the last few decades have been astonishing and have led to unprecedented levels of sensitivities, such that work with subnanomolar concentrations are now routine, while femtomolar and even single molecule measurements are possible with some effort. Although the "point and click" approach common with modern computer-interfaced instruments allows the novice to immediately carry out fluorescence measurements, an appreciation of the fundamental aspects of instrumentation is still indispensable. Such knowledge will absolutely help the beginner to avoid potential pitfalls and instrumental artifacts. Hence, in this chapter, I shall briefly review the properties of basic components of a modern spectrofluorimeter, namely: (1) light sources, (2) monochromators/filters, (3) polarizers, and (4) photodetectors (**Figure 3.1**). A word about nomenclature: when Enrique Gaviola described his instrument for measuring fluorescence lifetimes in 1926, he named it a *fluorometer*. For this reason lifetime instruments are referred to as *fluorometers*, while steady-state instruments are termed *fluorimeters*.

A schematic of a modern spectrofluorimeter is given in **Figure 3.2**, specifically, the instrument depicted is an ISS PC1 (ISS, Champaign, IL). We should note that virtually all modern spectrofluorimeters utilize a 90° angle between the excitation and detection directions, unlike spectrophotometers, which utilize a straight-through lightpath. The reason for this right-angle geometry is mainly to reduce the amount of exciting light in the observation direction. Ideally, one would like to observe the fluorescence against a dark background, devoid of any excitation photons. Although wavelength selection devices such

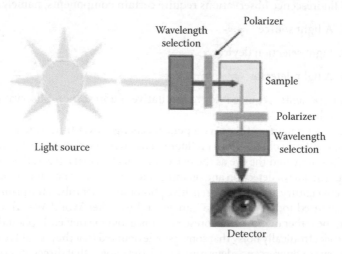

FIGURE 3.1 Basic components of a spectrofluorimeter.

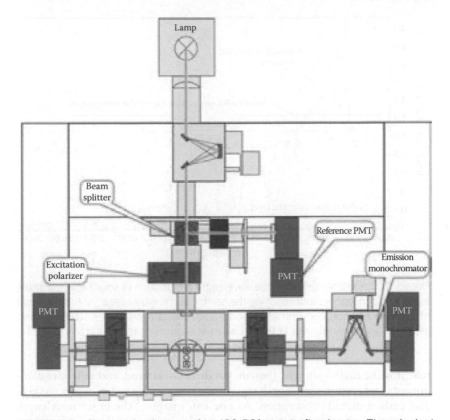

FIGURE 3.2 Schematic diagram of an ISS PC1 spectrofluorimeter. The principal components include the xenon arc lamp, the excitation and emission monochromators, a quartz bean splitter, a quantum counting solution (located between the beam splitter and the reference PMT), three PMTs, and excitation (P_{ex}) and emission (P_{em}) calcite prism polarizers. (The author would like to thank Beniamino Barbieri from ISS, Inc. for this figure.)

as filters or monochromators are used to isolate the emission wavelengths, these devices are not perfect, and in-line observation would put an intolerable strain on their effectiveness.

Light Sources

The earliest light source used for excitation of fluorescence was sunlight. Indeed, many of the seminal observations made by early investigators were made using sunlight admitted into a darkened room through a hole in a screen (George Stokes actually experimented with diverse sources of illumination including candles, gas flames, and even lightning). The spectral distribution of sunlight is depicted in **Figure 3.3**. Solar radiation from the sun corresponds roughly to blackbody radiation from a source near 5800°C. The radiation

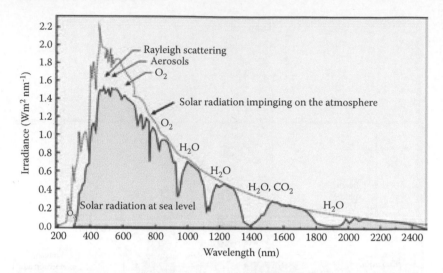

FIGURE 3.3 Sketch illustrating the wavelength distribution of solar radiation at the top of our atmosphere and reaching the earth's surface (sea level). (Modified from http://www.newport.com/Introduction-to-Solar-Radiation/411919/1033/content. aspx. The author would like to thank Leonel Malacrida for redrawing this figure.)

reaching the earth's surface, though, has direct, scattered, and reflective components and certain wavelength regions are removed by water, ozone, oxygen, and carbon dioxide absorption. As one can imagine, the exact wavelength composition of solar radiation on the earth's surface will also depend on latitude, longitude, and elevation. The wavelength regions between 315–400, 280–315, and 280–100 nm are known as the UVA, UVB, and UVC regions, respectively. Fortunately for life as we know it, the Earth's atmosphere, in particular, the ozone component, absorbs most of the UVC region.

I shall restrict this discussion to the light sources most commonly used in modern commercial spectrofluorimeters. These sources include xenon arc lamps, lasers, and light-emitting diodes (LEDs). Note that the light sources most commonly used in absorption spectroscopy, that is, in UV–Vis spectrophotometers, namely, deuterium and tungsten or halogen lamps, are rarely used in fluorescence, since they are relatively weak sources of photons. In addition to its intrinsic intensity, that is, the number of photons generated, the most important aspect of a light source for fluorescence measurements is the wavelength range available. From this point of view, the most popular light source for fluorescence instruments is the xenon arc lamp, introduced in 1951 by the Osram Company, which provides illumination from the ultraviolet to the infrared. This wavelength range is suitable for most biological samples since illumination of such samples is typically limited by absorption of water in the far-infrared, and by light-induced photochemistry in the deep-ultraviolet. The light distribution of a typical xenon-arc lamp, over the wavelength range most commonly used, is illustrated at the top of **Figure 3.4**. (*Note:* this

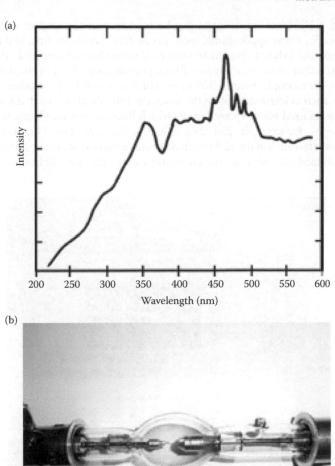

(a)

Intensity

200   250   300   350   400   450   500   550   600

Wavelength (nm)

(b)

FIGURE 3.4 (a) Spectral distribution from a xenon arc lamp using the ISS PC1 depicted in Figure 3.3. (Adapted from D.M. Jameson, J.C. Croney, and P.D. Moens, 2003. *Methods Enzymol.* 360: 1–43.) (b) Photo of 450 W xenon arc lamp.

spectrum is subject to the response characteristics of the monochromator utilized for scanning—more on this later.)

The bottom part of **Figure 3.4** shows a xenon arc bulb (I should note that other lamp designs are also available, such as Cermax® lamps with encased parabolic reflectors to increase output). These lamps use tungsten electrodes and contain xenon gas at pressures up to 25 atm. Ozone is produced by the action of UV light on oxygen, and so to prevent production of ozone in the air outside the lamp

31

housing, a UV-blocking material can be used to coat the interior of the bulb envelope. For many applications, such "ozone-free" lamps are fine, and have the advantage that exhaust systems to vent away ozone are not required. The catch, however, is that ozone-free lamps will not provide much light in the deeper UV regions—for example, below ~300 nm—which is necessary for studies on some systems, such as intrinsic protein fluorescence. I should also note that some early instruments used mercury arc lamps, which have intense emissions at specific wavelengths, for example, 254, 297, 302, 313, 365, 405, 436, 546, and 578 nm among others. Much of the early work on proteins carried out in the former Soviet Union utilized 297 nm, since mercury lamps were a common light source in that

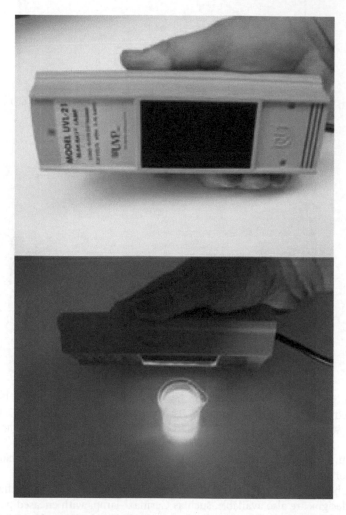

FIGURE 3.5 Top panel shows a UV handlamp. Bottom panel shows a solution of ANS in ethanol illuminated using the long wavelength (365 nm) selection of the handlamp.

country at that time. This wavelength was particularly useful for protein studies since, as will be discussed later, tyrosine residues have negligible absorption at this wavelength and tryptophan residues can be preferentially excited. The popular UV-handlamp (**Figure 3.5**) in fact utilizes two of the stronger mercury lines, 254 and 365 nm, for its "short" and "long" wavelengths, respectively. UV lamps were originally developed by Robert W. Wood, the famous American physicist (mentioned again later in this chapter), and even today are sometimes known as "Woods lamps" or "Black Lights." In the bottom half of **Figure 3.5** the "long" wavelength, that is, 365 nm, is shown illuminating a solution of ANS in ethanol.

In the last couple of decades, LEDs and lasers, specifically laser diodes (LDs), began to appear as light sources in basic spectrofluorimeters. Lasers have, of course, been around since the 1950s (the reader should know that the word "laser" is an acronym for "Light Amplification by Stimulated Emission of Radiation") but for decades were large and expensive devices. Since the more recent development of LDs, and their utilization in the telecommunication industry as coupled light sources for fiber optics communications as well as barcode readers and optical recording devices (such as CD readers and writers), these devices have become much less expensive and hence more widely utilized. Some of the wavelengths emitted by commonly used lasers are indicated in **Figure 3.6**. In recent years, a new laser source called a "white laser" or a "super continuum laser" has appeared. The output of this laser covers a wide range of wavelengths, typically from 400 to 1000 nm (**Figure 3.7**). Obviously, if a white laser is used as an excitation source for fluorescence experiments, a wavelength selection device, that is, a monochromator or filter, must be used to isolate the wavelength range desired. The advantages of the white laser source are high intensity, sharp directionality, and wide wavelength range. A disadvantage is that light below about 400 nm is not available at present—and these devices are also very expensive!

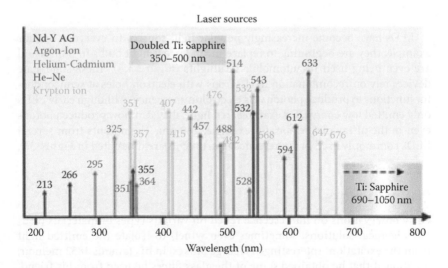

FIGURE 3.6 Depiction of wavelengths available from diverse lasers.

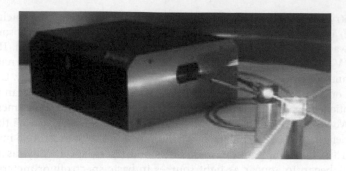

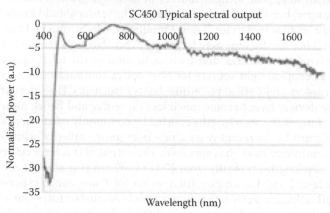

SC450 Typical spectral output

FIGURE 3.7 Fianium SC450 Supercontinuum laser and its wavelength output. (Downloaded on August 7, 2013 from http://forc-photonics.ru/data/files/sc450-pp.pdf.)

LEDs have become increasingly popular light sources in everyday life, for example, they are beginning to replace incandescent light bulbs in homes and are even being used in automobile headlights (**Figure 3.8**). These solid state devices rely on recombination of electrons with electron holes at semiconductor junctions to produce photons via electroluminescence. Although early LEDs only emitted low energy, for example, red light, they can now produce photons even in the ultraviolet region, for example, 260 nm. The outputs from several LEDs commonly used in fluorescence instruments are illustrated in **Figure 3.9**.

Optical Filters

Early observations on fluorescence relied on various types of materials (usually chemical solutions, sometimes even wine!) to isolate the emitted light from the excitation. Interestingly, George Stokes, in his famous 1852 memoir, mentioned that he obtained some of the glass filters he used from his friend, Michael Faraday! Filters are still used today in many applications, including

FIGURE 3.8 Illustration of several LEDs. (Downloaded on August 7, 2013 from http://www.ledmodulemanufacturer.com/supplier-led_light_emitting_diode-1549.html.)

routine intensity measurements as well as polarization and lifetime measurements and, of course, microscopy. Many filters used today are made of sophisticated glasses or plastics, and include longpass (also known as cut-on, and formerly known as cut-off), bandpass, and interference filters. Longpass filters

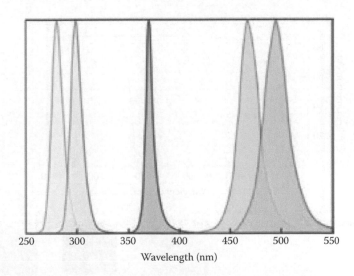

FIGURE 3.9 Output from different LEDs. Nominal wavelengths, from left to right, are 280, 300, 370, 471, and 488 nm. (Modified from ISS Inc., Champaign, IL. The author would like to thank Leonel Malacrida for redrawing this figure.)

are designed to strongly block all wavelengths below a certain range but to allow wavelengths above that range, that is, at longer wavelengths, to pass. Examples of such longpass filters are shown in **Figure 3.10** (figure courtesy of Theodore Hazlett). When I was a graduate student with Gregorio Weber, we routinely used a thin (2 mm) solution of 2 M sodium nitrite as a longpass filter, specifically, to block wavelengths below about 400 nm. To be more specific, we would use the $NaNO_2$ filter to block excitation from hitting any emission filter used in tandem, for example, a yellow filter used to block blue wavelengths. Some of the longpass filters available at that time would give off some fluorescence/phosphorescence and the $NaNO_2$ filter was intended to block most of the exciting light from hitting the longpass filter, thus greatly reducing spurious emissions, especially when turbid samples were being studied. Nowadays, incomparably better "low-fluorescence" longpass filters are available.

Bandpass filters are designed to pass a given range of wavelengths—for example, a 50–150 nm range—but to strongly block wavelengths outside of this window. Interference filters, like bandpass filters, are designed to pass only a given range of wavelengths, and are designed using dielectric coatings, which give rise to constructive and destructive interference of light, to pass only a narrow range of wavelength, typically 10 or 20 nm. The terms "bandpass" or "interference" filters are perhaps outdated, since now filters are available which can pass an arbitrary range of wavelengths. In 2000, the company Semrock was founded and, using state-of-the-art ion-beam sputtering technology that allowed the use of hard dielectric coatings on a single glass substrate, produced

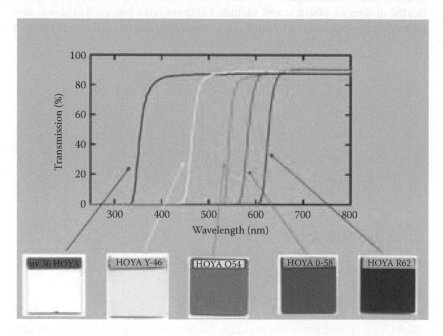

FIGURE 3.10 Depiction of five longpass filters and their associated transmission spectra. (The author would like to thank Theodore Hazlett for this figure.)

a new generation of thin-film interference filters that have exceptional optical qualities, in particular very high transmissions. Transmission curves for two of these bandpass/interference filters are shown in **Figure 3.11**. A more recent development is the tunable interference filter, which allows one to select different narrow wavelength regions by adjusting the angle between the filter and the incident radiation (**Figure 3.11**). Another filter type that I use quite often, for both steady-state and time-resolved measurements, are neutral density filters.

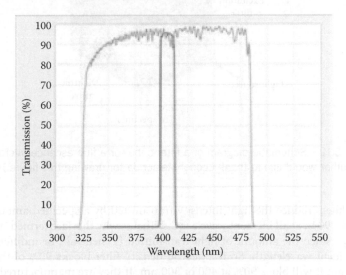

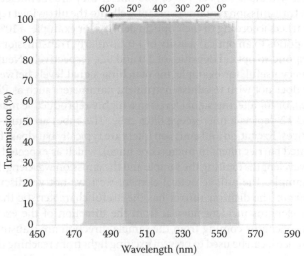

FIGURE 3.11 Transmission spectra for two bandpass filters and a tunable filter. Both bandpass filters are centered at 405 nm, with FWHM of 150 nm (green) and 10 nm (red). (The author would like to thank Jim Passalugo of Semrock, Inc. for these spectra.)

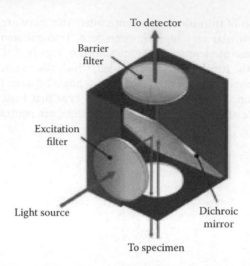

To detector

Barrier
filter

Excitation
filter

Light source

Dichroic
mirror

To specimen

FIGURE 3.12 Schematic diagram of a filter cube for a fluorescence microscope. (The author would like to thank Leonel Malacrida for drawing this figure.)

These filters reduce the light intensity transmitted by a specified amount, for example, 90%, at all incident wavelengths. That is why they are termed "neutral density," since the amount of light transmitted is "neutral"; or indifferent, to the incident wavelength. So if a neutral density filter blocks 90% of the light at 500 nm it will block 90% at 400 or 300 nm. If they are manufactured out of quartz, the transmission remains constant well into the ultraviolet region. Often they are marked according to their optical density, for example, a 10% transmission corresponds to an optical density of 1.0. Similarly, transmissions of 1% and 50% correspond to optical densities of 2.0 and 0.3, respectively. Neutral density filters are very useful, for example, for matching signal levels between samples (when one does not wish to adjust instrument parameters such as PMT voltage or slits, etc.), or for altering signal levels in a highly reproducible manner.

Figure 3.12 depicts a microscope filter cube. The cubes are built using three optical devices. Excitation and emission filters are typically bandpass filters (commonly termed barrier filters in microscope lingo), which as expected are used to select the wavelengths exciting the sample and the emission wavelengths collected from the sample. The unique optical element we have not yet discussed is the *dichroic mirror*. The dichroic mirror has the useful characteristic that it reflects certain wavelengths incident upon it from the direction of the excitation, but different wavelengths coming from the sample direction are transmitted to the detector. Hence it can be used to prevent exciting light from reaching the detector.

Monochromators

People had experimented with prisms and light before Isaac Newton—but it was generally believed that the prism somehow "colored" the light. Newton

FIGURE 3.13 Depiction of a prism dispersing light. (The author would like to thank Norin Redes from ISS, Inc. for this figure.)

was the first to clearly understand that the prism revealed an underlying characteristic of white light—namely that it was composed of many colors (**Figure 3.13**).

This realization essentially gave birth to monochromators. Early monochromators used prisms to disperse the incident light. The problem with monochromators based on prisms, however, is that the light dispersion is not linear with wavelength and also that normal glass prisms do not pass UV light (below about 340 nm)—so expensive quartz prisms must be used if ultraviolet light is required. For these reasons, grating-based systems became more popular. The diffraction grating was first introduced by the American astronomer David Rittenhouse in 1785, but was popularized in 1813 by Joseph von Fraunhofer, who first resolved the sodium D lines. Although his first diffraction gratings consisted of fine parallel wires, Fraunhofer soon found that etching—or ruling—grooves in metal surfaces could achieve the same result; his first surface was glass coated with gold. Fraunhofer was soon able to use his gratings to resolve the lines in the solar spectrum that still bear his name. Henry A. Rowland (1848–1901), a Professor at Johns Hopkins University, created sophisticated "ruling engines" that produced a series of ruled diffraction gratings, which set the standard for many years. In the late 1960s, however, optical methods such as photographic lithography and holography began to be used to produce very high-quality diffraction grating. Diffraction gratings operate using the principle of interference, specifically constructive and destructive interference of light, as illustrated in **Figure 3.14**. A diffraction grating in operation in an ISS spectrofluorimeter is shown in **Figure 3.15** (photo courtesy of Theodore Hazlett). The monochromator shown, like most monochromators used in spectrofluorimeters, is a Czerny–Turner design. Although monochromators nominally pass a given wavelength of light, it is important to realize that the wavelength selected is not completely specific. In other words, at a given monochromator setting, a range of wavelengths are passed as determined by the physical width of the slits (**Figure 3.16**). Clearly, the wider the slit the larger the range of wavelengths selected. This range is usually quantified by the full-width (in nanometers) of light passed

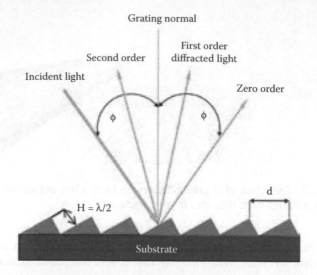

FIGURE 3.14   Depiction of a ruled diffraction grating.

FIGURE 3.15   Inside of a monochromator.

FIGURE 3.16 Three slits with different fixed widths: from left to right, 0.5, 1.0, and 2.0 mm. The dispersion of the monochromator is 8 nm/mm.

at half-maximum intensity, or FWHM. The relationship between the FWHM and the slit width is given by the *dispersion* of the monochromator. For example, if the dispersion of the monochromator is 8 nm/mm then slits of 2 mm will result in a bandpass of 16 nm FWHM. In this case, if the center wavelength is set to 500 nm, then half of the intensity passed at 500 nm will be passed at 508 nm and 492 nm. Of course, in principle, one could use smaller slits to get narrower FWHMs and hence more monochromatic light. But the tradeoff is intensity, since each time a slit width is halved, the intensity at the maximum is also halved. Some monochromators used fixed slits, such as those shown in **Figure 3.16**, while others use a variable width design, that is, a mechanism that allows the slit width to be adjusted to any value in a particular range. Each design, fixed or variable, offers its own advantage, for example, the fixed slit design allows for excellent reproducibility of conditions, whereas the variable slit design allows for rapid, "on the fly" changes and more flexibility in the choice of slit width.

Monochromators are not ideal wavelength selection devices since their throughput efficiencies vary with both the wavelength as well as with the polarization of the light (polarization will be discussed in Chapter 5). One of the most dramatic examples of the nonideality of monochromators is the famous Wood's Anomaly, named after Robert W. Wood who discovered the effect in 1902. A theoretical description of this phenomenon is beyond the scope of this book—and certainly beyond the scope of the author! But in practical terms what happens is that at some wavelengths, much of the energy of the component of the light that is polarized perpendicular to the grooves on the diffraction grating is lost with the result that a relatively sharp drop in the light intensity occurs at that wavelength. This energy loss does not occur in the parallel polarized light

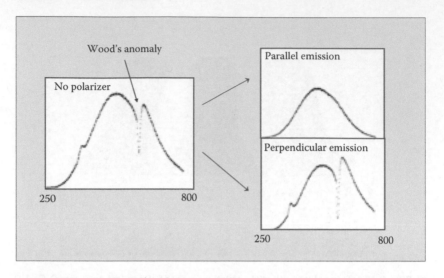

FIGURE 3.17 Depiction of Wood's anomaly in an SLM monochromator.

component. This effect is shown in **Figure 3.17**. This effect will be discussed again in the next chapter (and makes the Rogue's Gallery list in the Appendix). *Anecdote alert! I was once teaching at a workshop in Australia and was showing students the Wood's anomaly on the instrument provided. To my consternation, it did not disappear when I rotated the emission polarizer to the parallel (vertical) orientation relative to the laboratory vertical axis—in fact, it was more pronounced. I thought a moment and then said to the company rep, who was hovering nearby, "Let me guess, in this instrument you mount the monochromators at 90° relative to the vertical axis don't you?" He replied, "Yes—how did you know?" The take home lesson is that elimination of the Wood's anomaly requires viewing the emission through a polarizer oriented parallel to the direction of the lines on the diffraction grating—which is usually, but not always—parallel to the laboratory vertical axis.*

Since diffraction gratings work on the principle of interference, one should realize that a given wavelength setting may not pass exclusively a unique wavelength of light. For example, if a monochromator is set at 600 nm, it will pass 600 nm light, but it will also pass 300 nm light. In fact, if one looks inside a monochromator while light is incident upon the diffraction grating (as in **Figure 3.15**), one can see the so-called higher orders of light on the interior walls. Another aspect to consider is that monochromators (like book authors), are not perfect and will always transmit a small amount of light at wavelengths other than the wavelength they are set at. This light is known as "stray light" or "parasitic light" and will vary depending on the monochromator design and the quality of the diffraction grating. Gratings prepared by holography, for example, have much less errors in the spacings and depths of their grooves compared to mechanically ruled gratings, and typically produce much less stray light. Peaks due to stray light are sometimes referred to as Rayleigh ghosts, and can be problematic when the sample is turbid, for example, a membrane preparation, since these ghosts can then be more readily scattered toward the detector. An interference filter, optimized for the selected excitation wavelength, placed in the light path after the excitation monochromator, can be used to exorcise these Rayleigh ghosts.

Polarizers

In 1669, Erasmus Bartholin discovered a curious optical phenomenon, namely, double refraction of light, in a crystalline material known as Iceland spar. In 1808, Etienne-Louis Malus observed sunlight reflected from the windows of the Luxemburg Palace in Paris through an Iceland spar crystal that he rotated— we now know that Iceland spar is composed of the mineral known as calcite. Malus discovered that the intensity of the reflected light varied as he rotated the crystal and coined the term *"polarized"* to describe this property of light. Malus also derived an expression for calculating the transmission of light as a function of the angle (θ) between two polarizers. This equation (Malus' law) is now written as: $I_\theta = I_0 (\cos^2\theta)$. David Brewster studied the relationship between refractive index and angle of incidence on the polarization of the reflected light. He discovered that for normal glass and visible light, an incidence angle of ~56° resulted in total reflection of one plane of polarization—this angle is now known as *Brewster's Angle* (θ_B) and in general:

$$\theta_B = \tan^{-1} [\eta_2/\eta_1]$$

This discovery allowed Brewster to construct a polarizer composed of a "pile-of-plates" (**Figure 3.18**). In fact, I found such a pile-of-plates polarizer in a cabinet in Gregorio Weber's lab when I was a student, and he explained its function to me and said that he had used it during his thesis work!

In 1828, William Nicol joined two crystals of Iceland spar, cut at an angle of 68°, using Canada balsam, which allowed a spatial separation between the orthogonally polarized components (**Figure 3.19**). Other important calcite polarizers developed around this time include: Glan–Foucault, Glan–Thompson, Glan–Taylor, Wollaston, and Rochon. Most modern spectrofluorimeters today use Glan–Taylor-type calcite polarizers, which have an air-gap between the two calcite crystals allowing for transmission deep into the ultraviolet. The Henry Ford of polarizers however, was Edwin Herbert Land. In 1929, Land patented the sheet polarizer (the J-sheet), consisting of crystals of iodoquinine sulfate embedded in nitrocellulose film followed by alignment of the crystals by stretching, which led to dichroism. In 1937, he founded the Polaroid Company and in 1938, he invented the H-sheet which was comprised of polyvinyl alcohol sheets with embedded iodine. Land's invention made him quite rich and also allowed the use of inexpensive polarizers in diverse applications such as sunglasses and photography filters. A few modern instruments use such sheet polarizers instead of the more expensive

(a)

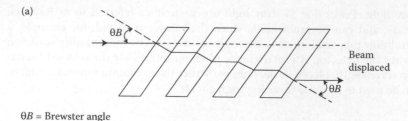

θB = Brewster angle

(b)

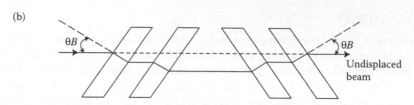

FIGURE 3.18 Depiction of two designs of pile-of-plates polarizers. (a) Linear stacked array and (b) opposed array.

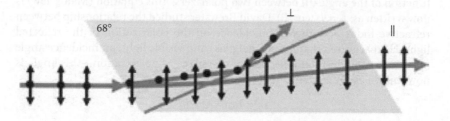

FIGURE 3.19 Schematic of a calcite polarizer.

calcite polarizers. For many applications this approach is fine, except that such polarizing sheets may be opaque in the ultraviolet and may also show decreasing ability to isolate the parallel and perpendicular components from one another. As usual, the tradeoff is between cost and performance. Several types of polarizers are shown in **Figure 3.20**.

Detectors

As mentioned before, the eye was certainly the principal detector used in early fluorescence investigations. And visual observation is still a very useful approach! In fact, when conducting workshops on fluorescence, I always tell the students what Gregorio Weber used to tell me—namely, if the fluorescence lies in the visible wavelength region, always look at the sample in the instrument. Not only will you enjoy the colors, but you may be able to tell if everything looks correct (more on this later)! Although the human eye is a wonderful photodetector, it does not easily interface with computers, and

FIGURE 3.20 Photos of three types of polarizers: from left to right (a) H-sheet polarizers, (b) a Glan–Taylor calcite crystal polarizer, and (c) the author's Maui Jim polarizing sunglasses.

its useful wavelength range is limited, covering only a narrow region of the electromagnetic spectrum, around 380–720 nm. The Young–Helmholtz theory, developed in the nineteenth century by Thomas Young and Hermann von Helmholtz, proposed the idea that color vision is trichromatic, due to the three retinal visual pigments in the retina's cone cells, which respond primarily to blue (short wavelength), green (medium wavelength), and red (long wavelength) light (**Figure 3.21**). Genes coding for medium-wavelength and long-wavelength photopigments are X chromosome inherited, and hence color vision is a sex-linked trait. Fairly recently, it was discovered that some females have tetrachromatic vision (like certain fish, birds, and insects), that is, they have an additional cone cell that specifically detects yellow wavelengths, in between the green and red cone cells. Given the inherent limitations of the human eye (and to avoid endless arguments between male and female observers), it is convenient to use an electronic device—such as a photomultiplier tube (PMT) for photodetection. The modern PMT takes advantage of the photoelectric effect, discovered in 1887 by Heinrich Hertz. The basic idea is that

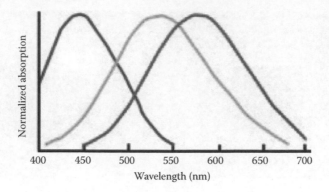

FIGURE 3.21 Wavelength absorption characteristics of the cone cells in the human eye.

photons impacting on a suitable material can cause the release of electrons (termed photoelectrons). These electrons can be detected and hence give a measurement of the intensity of incident light. To greatly improve upon the sensitivity of such photodetectors, one can set up a series of photosensitive plates—called dynodes—having a net positive charge relative to the previous plate—such that photoelectrons will be attracted to the plates and give rise to more photoelectrons. This principle is illustrated in **Figure 3.22**.

These devices are termed photomultipliers, since the effect of one incident photon is greatly multiplied by the subsequent stages. Modern PMTs are constructed with multiple stages (typically 10–14) and may be either end-on or side-on as illustrated in **Figure 3.23**.

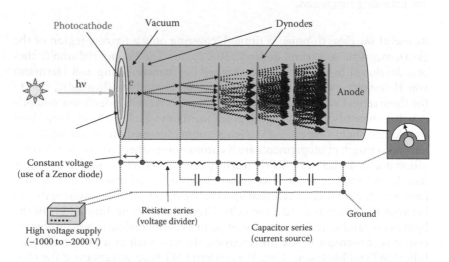

FIGURE 3.22 Schematic of an end-on PMT.

FIGURE 3.23   Photo of side-on and end-on PMTs.

Photon Counting versus Analog Detection

Many commercial spectrofluorimeters utilize analog detection, which means that the photocurrent from the PMT is converted into a voltage, which can then be converted into a digital signal (using an analog-to-digital converter), appropriate for interface with a computer. Another popular method uses the photon counting approach. This approach is illustrated in **Figure 3.24**. In essence, the photocurrent from the PMT is amplified and passed to an electronic device known as a discriminator. The discriminator will produce a clean output pulse whenever the photocurrent exceeds a given threshold. In this way, the original analog signal is directly transformed into a digital signal. Each of these approaches, analog and photon counting, have their characteristic advantages and drawbacks. The photon counting method offers the best signal-to-noise ratio when the signal is weak, but it cannot be used for very large signals. The upper limit has to do with the width of the pulses generated. Essentially, if another photon arrives during the interval when an earlier photon is being processed, the system cannot "count" the second photon, and hence the phenomenon known as "pulse pile up" occurs, and the count rate will not be linearly related to the incident signal. The analog method, in comparison, is excellent when one is working with a signal which encompasses a large dynamic range, since one can readily raise or lower the high voltage to the PMT to adjust the signal. At low light levels, however, the signal-to-noise ratio can be poor.

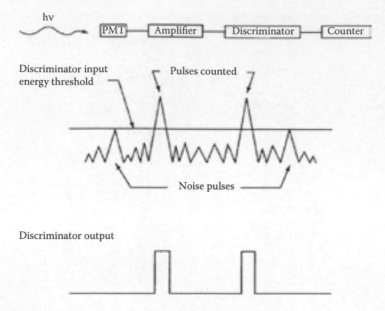

FIGURE 3.24 Depiction of the photon-counting technique.

Just as the human eye has an effective detection range, photodetectors also have upper and lower detection limits. When I began as a graduate student in the early 1970s, the PMTs we used in Weber's lab were only useful up to around 600 nm or so. I remember when Weber came into the lab one day and handed me a PMT he had received as gift from the Hamamatsu Corporation. He told me that it had more red-sensitivity than the PMTs we were presently using and asked me to try it out. After some days of constructing a suitable housing, I installed it on our spectrofluorimeter and was amazed at how far out into the red region it would detect light. That tube, now known as an R928, became one of the most popular PMTs ever used and certainly greatly facilitated the determination of "corrected" emission spectra (more on this topic later). The response characteristics of the R928 PMT are shown in **Figure 3.25**.

An increasingly popular detector, especially for certain microscopy applications, is the APD or avalanche photodiode. This solid state device, originally developed during the 1960s and 1970s, is essentially the semiconductor equivalent of the PMT. The wavelength detection range, sensitivity, and speed of an APD depends upon the precise makeup of the semiconductor material, that is, silicon, germanium, gallium, and so on. APDs that cover the wavelength range from ~200 to ~1700 nm are presently available. They are generally small (active surface areas in the range of hundreds of square microns), typically have better signal-to-noise characteristics in the redder regions of the spectrum (e.g., >500 nm) than PMTs and operate at significantly lower voltages than PMTs. Currently, however, they are generally more expensive than most PMTs. A schematic diagram of an APD is shown in **Figure 3.26**.

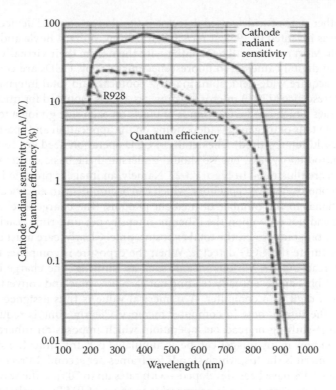

FIGURE 3.25 Wavelength response of a R928 PMT.

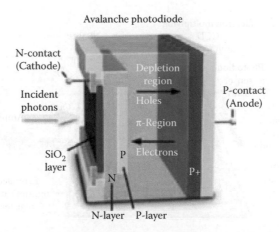

FIGURE 3.26 Schematic diagram of an Avalanche photodiode. (The author would like to thank Michael Davidson and the Hamamatsu Learning Center (http://learn. hamamatsu.com/) for this figure.)

Another type of widely used detector is the charge-coupled device (CCD), which was invented in 1969 at AT&T Bell Labs by Willard Boyle and George E. Smith. Most readers will be familiar with them from their virtually universal use in digital cameras. In fluorescence microscopy, CCDs are commonly used to acquire wide-field, spinning-disk confocal, and total internal reflection fluorescence (TIRF) microscopy images. CCDs mounted for microscopy have arrays which can be thousands of pixels on each side giving a total area of a few to tens of megapixels (depending on the application and spatial–temporal resolution required). The details of CCD operation need not concern us in an introductory text, but we should understand the basic concepts, some of which are illustrated in **Figure 3.27**. Namely, an image is projected onto an array of photosensitive capacitors, which can accumulate an electrical charge proportional to the number of incident photons. The charge level attained will depend not only upon the inherent intensity per unit time reaching the capacitor bin, but also on the total exposure time. Clearly, care must be taken not to saturate the CCD detector. When the exposure is complete, the data can be "read-out" by various means, such as "shifting" the charge between capacitor bins and eventually reading out each bin value and converting that value to a digit in a computer. A numerical value is thus assigned to each pixel in the image, now in computer memory. Clearly, time is required for the charge-shifting or read-out operation, which imposes an inherent constraint on how many images can be accumulated per unit time: the standard figure of merit is the fps, which stands for frames per second. In recent years, very fast CCDs have been developed—with rates up to 70 fps. The wavelength response of modern CCDs is comparable to those of PMTs. A relatively new method of increasing sensitivity involves electron multiplying CCD technology (EMCCD), which utilizes an on-chip multiplication gain. The operational principles of an EMCCD are shown in **Figure 3.27**.

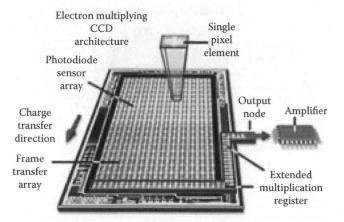

FIGURE 3.27 Schematic diagram of a charge-coupled device. (The author would like to thank Michael Davidson and the Hamamatsu Learning Center (http://learn. hamamatsu.com/) for this figure.)

Miscellaneous Topics

Cuvettes

The majority of *in vitro* fluorescence measurements are carried out using fluorescence cuvettes (*Note*: "cuvette" is occasionally spelled "cuvet," but the longer form is more common). A glance at any web site featuring cuvettes will show a myriad of options differing in materials and pathlengths or geometries. Here I shall illustrate only the most commonly used cuvettes.

On the off-chance that some readers have never seen such devices, I include some pictures. **Figure 3.28** shows two standard 1 cm pathlength cuvettes—on the left a fluorescence cuvette and on the right an absorption cuvette. The difference is that the absorption cuvette is polished only on two opposing sides while the fluorescence cuvette has all four sides polished. The reason for this difference is, of course, the fact that it costs money to polish cuvettes, and since absorption studies are carried out on a straight-through light pathway, only two opposing polished sides are required. As previously mentioned, fluorescence measurements using cuvettes are almost always realized using a right angle geometry so four sides are polished (I suppose one could get away with three sides polished but I have never seen such an arrangement, and one would inevitably find the wrong orientation being used)! Surprisingly, the use of an absorption cuvette instead of a fluorescence cuvette does not necessarily

FIGURE 3.28 Photo of a fluorescence cuvette (left) and absorption cuvette (right).

reduce the intensity as much as one might suppose—only about 10% in the instrument I tested—but it certainly is not an aesthetically pleasing way to do measurements and it would sound very silly when you write the Methods part of the manuscript! *Anecdote alert! When I was a graduate student, I was trying to improve the sensitivity of my measurements and I hit upon the idea of having two adjacent sides of a fluorescence cuvette coated with a mirror finish. My idea was that the excitation beam would then be reflected from the back side through the solution again, and the fluorescence reaching the side facing away from the detector would be reflected toward the detector. In fact, this arrangement improved my signal about three-fold. I was, naturally, proud of this accomplishment and demonstrated it to Gregorio Weber who politely praised my ingenuity and then proceeded to show me an old article he had written (from the 1950s) in which he had also used mirror coatings on his cuvette. In that article, he also acknowledged that he was following the idea of Francis Perrin who published the same approach in the 1920s!*

Cuvettes may be manufactured using different materials, namely different types of glass or quartz or even plastic. The choice of cuvette depends on the application, of course, and may be dictated by considerations of cost. Glass cuvettes transmit wavelengths down to about 330 nm or so, depending on exactly the type of glass used. Quartz will transmit well into the ultraviolet; 200 nm can be reached with high-quality quartz cuvettes. Some quartz has endogenous fluorescence, however, and it is best to use Suprasil quartz, which exhibits very low fluorescence. Disposable cuvettes, usually made of some type of plastic, are much less expensive than glass or quartz cuvettes but do not transmit as far in the UV. Some plastic cuvettes only transmit down to about 350 nm but newer types of plastic are available, which can transmit light well below 300 nm. The point here is that you should consider your application and be certain that your cuvettes will allow you to work in the necessary wavelength range.

Another important issue is the pathlength. It is traditional to use 1 cm pathlength cuvettes and indeed almost all sample compartments are machined to hold a 1 cm square cell. If one uses a 1 cm square cuvette, the lowest sample volume which can be used is typically about 2 mL, which should be checked in the particular instrument utilized (in some instruments the light passes through the center of the cuvette while in others it hits close to the bottom). In many cases, though, samples can be precious and less volume is desirable. In these cases, one can use smaller pathlength cuvettes, which can either be the so-called T-format, shown in **Figure 3.29**, or a smaller square cuvette (which must be placed in an appropriate adapter to fit in the sample compartment), shown in **Figure 3.30**, along with a triangular cuvette for front-face excitation. Front-face cuvettes are useful if one must look at a sample of high concentration, such that the optical density at the excitation wavelength would cause severe inner filter effects in normal cuvettes. The challenge with using traditional front-face excitation, however, is the risk of spurious scattered light. This problem arises from the fact that the 45° angle of the cuvette face will directly reflect excitation light toward the detector. This "specular" reflection puts a serious burden on the wavelength isolation device on the emission side,

FIGURE 3.29 Photos of dual pathlength fluorescence cuvettes. Front and side views.

FIGURE 3.30 Photos of three types of fluorescence cuvettes: from left to right standard 10 mm pathlength, 3 mm pathlength, and triangular cuvette. Front and top views.

be it a filter or monochromator. Also, any parasitic light from the excitation system, that is, the Rayleigh ghosts, will be reflected directly to the emission arm. For front-face excitation, it is better if one can use an angle other than 45° for the reflecting surface, for example, 30°. This concept is illustrated in **Figure 3.31**. The figure shows a concentrated solution of rhodamine in ethanol being excited with green (500 nm) light; the normal sample holder has been removed to allow for facile orientation of the cuvette relative to the

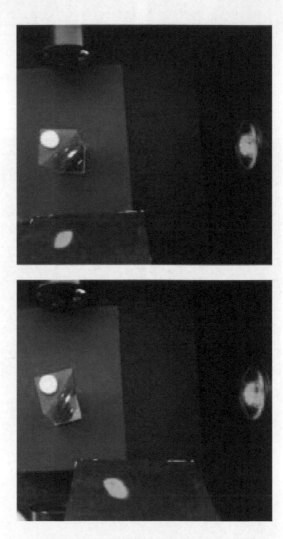

FIGURE 3.31 Triangular cuvette containing concentrated solution of rhodamine in ethanol; sample compartment viewed from above. (Top) Cuvette oriented such that the specular reflection enters the observation optical path. (Bottom) Orientation set such that specular reflection is angled away from the observation optical path.

FIGURE 3.32 Top view of a variable angle front face sample holder.

exciting light. In the top photo, the triangular cuvette is in the normal orientation, that is, as it would sit in a conventional sample holder. One can see that the specular reflection, the green light, is being reflected at 90° to the excitation direction and would thus enter the collection lens. When the cuvette is rotated counterclockwise, however, the specular reflection moves outside of the observation volume of the lens, though the fluorescence stays stationary and can still be observed. To facilitate the process of sample orientation, some instruments can be provided with a variable angle sample holder as shown in **Figure 3.32**, which allows one to select the angle of the sample orientation relative to the excitation direction. This type of sample holder is particularly useful for solid samples. For example, I recently used this sample holder to help someone look at the fluorescence from various types of paper.

Finally, I should point out that a great variety of "specialty" cuvettes exist. For example, there are microcuvettes, which can handle extremely small volumes on the order of a few microliters, as well as cuvettes that allow for anaerobic measurements. There are also cuvettes adapted to act as flow-cells.

Plate Readers

The first microliter plate was developed in 1951 by Gyula Takátsy. Over the next 60 years, enormous progress was made in the design of plates and plate readers, and a wide variety of formats are now available. The classic 96-well

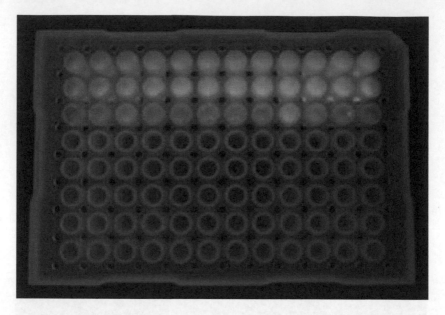

FIGURE 3.33 A 96-well plate with some wells containing fluorescent solutions.

format (8 × 12) is shown in **Figure 3.33**. Typical well capacity is 100–200 μL. Larger capacity plates are now commonplace including 384 and 1536 well formats. These systems were originally designed for cell cultures and cell assays but were quickly adapted for high-throughput screening, primarily in the drug discovery field. Although originally restricted to absorption measurements, plate readers were soon developed for fluorescence measurements, and presently plate readers are available which can not only record fluorescence intensities but also spectra, polarizations, and lifetimes.

Additional Reading

S. Udenfriend, 1995. Development of the spectrophotofluorometer and its commercialization. *Protein Sci.* 4: 542–551.

J.C. Croney, D.M. Jameson, and R. Learmonth, 2001. Teaching basic fluorescence principles with simple visual demonstrations. *J. Biochem. Ed.* 29: 60–65.

D.M. Jameson, J.C. Croney, and P.D. Moens, 2003. Fluorescence: Basic concepts, practical aspects and some anecdotes. *Methods Enzymol.* 360: 1–43.

S. Das, A.M. Powe, G.A. Baker, B. Valle, B. El-Zahab, H.O. Sintim, M. Lowry, 2012. Molecular fluorescence, phosphorescence, and chemiluminescence spectrometry. *Anal. Chem.* 84: 597–625.

Emission and Excitation Spectra

Emission Spectra

As already mentioned, the emission spectrum represents a plot of fluorescence intensity versus wavelength (or wavenumber). Such a plot is, in principle, easy to obtain. As with many endeavors, however, the devil is in the details! **Figure 4.1** shows the emission spectra for several commonly used fluorophores. The actual samples, illuminated with a UV handlamp, that give rise to the spectra are also shown, since I believe it is useful for one to be able to roughly gauge the emission maximum of a sample from the color. Certainly, a "trained eye" can be an advantage when one is planning to record an emission spectrum since it can provide an estimate of the range of wavelengths which should be scanned.

Why Do We Want Emission Spectra?

We may ask the question, "What information is available from an emission spectrum?" We can imagine a wide range of answers, encompassing qualitative to quantitative considerations. Some researchers may simply care to use such spectra in an analytical sense, for example, comparing intensities of diverse samples to judge how much fluorescent material is present, which may then be used to judge how much of a target analyte is present. This latter case covers the very broad and diverse field of sensors, for example, molecules designed to bind to or react with specific ions (calcium, zinc, lead, protons, etc.) or antigens, or other chemicals or biomolecules. In the case of some ion binding fluorophores, the binding process is accompanied by a distinct change in spectral shape. Another use for an emission spectrum is to gain information on the local environment surrounding the fluorophore. For example, some fluorophores demonstrate a red shift in their emission spectrum as the

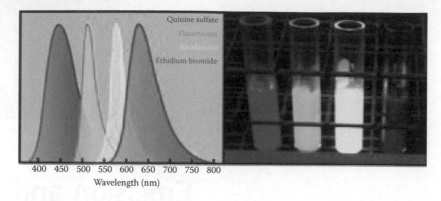

Quinine sulfate
Fluorescein
Rhodamine
Ethidium bromide

400 450 500 550 600 650 700 750 800
Wavelength (nm)

FIGURE 4.1 Emission spectra of quinine sulfate, fluorescein, rhodamine, and ethidium bromide (left). Solutions used for emission spectra (right).

polarity of their solvent environment increases. These types of environmentally sensitive probes will be discussed in more detail in Chapter 10.

Emission spectra may also be very important in FRET studies, which will be discussed in Chapter 8. Other specialized applications of emission spectra appear in studies of intrinsic protein fluorescence, the topic of Chapter 11. Suffice to say, the emission maximum of the intrinsic tryptophan fluorescence of a protein can be useful to detect and monitor conformational alterations in proteins, for example, in response to ligand binding, subunit association, or denaturation. Finally, one may be interested in the change in intensity of a fluorophore, for example, in response to a quencher molecule (more on quenching in Chapter 6), but instead of simply monitoring the intensity at a specific emission wavelength, one may need to integrate the entire spectrum using the area under the curve to judge the change in yield. This approach to quenching is often required if a wavelength shift accompanies the change in yield as often happens in the case of intrinsic protein fluorescence, for example, in chemical denaturation studies—these types of studies will be discussed in more detail in Chapter 11.

General Considerations

I have noted that in the last decade or so there seems to be a trend to simply record fluorescence intensities and to ignore spectra. Of course, if one is working with a microscope or a high-throughput multiwall plate reader, intensity measurements are sufficient for many purposes. Still, I urge the fluorescence practitioner working with *in vitro* systems to take emission spectra of their samples if at all possible, since that information may help one to avoid problems. A classic difficulty I have seen, for example, is the tendency to work at extremely low concentrations such that the "sample signal" is dominated by fluorescence and/or scatter from the background. Modern fluorescence instrumentation is so sensitive that one will almost always be able to find a "signal" no matter how dilute the sample!

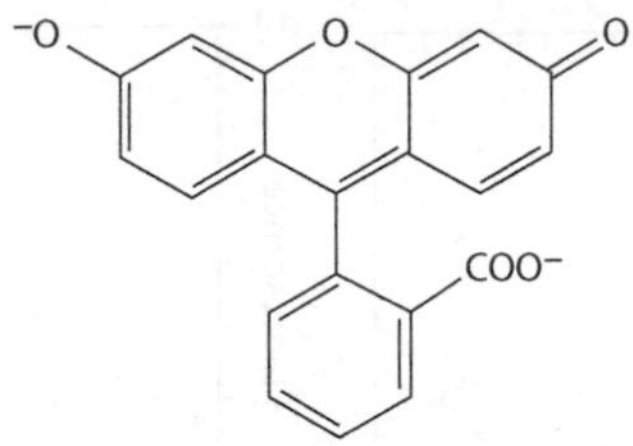

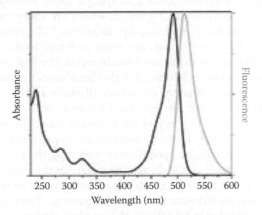

FIGURE 4.2 Structure of an anionic form of fluorescein.

Early examination of a large number of emission spectra resulted in the formulation of certain general guidelines, namely,

1. In a pure substance existing in solution in a unique form, the fluorescence spectrum is invariant, remaining the same independent of the excitation wavelength
2. The fluorescence spectrum lies at longer wavelengths than the absorption
3. The fluorescence spectrum is, to a good approximation, a mirror image of the absorption band of least frequency

Let us illustrate these points by considering a typical emission spectrum. It seems appropriate to use fluorescein for our example since it is certainly one of the most popular fluorophores used as well as being the first synthetic fluorophore (**Figure 4.2**). Note that the fluorescein structure depicted carries two negative charges, which corresponds to its ionic state at alkaline pH. The absorption and emission spectra of fluorescein in 0.01 N NaOH is shown in **Figure 4.3**. We may immediately note that the emission maximum is shifted to longer wavelengths, compared to the absorption spectrum.

FIGURE 4.3 Absorption and emission spectra of fluorescein in alkaline solution. (The author would like to thank Carissa Vetromile for this figure.)

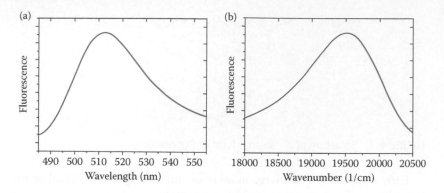

FIGURE 4.4 Emission spectra of fluorescein plotted against wavelength (a) and wavenumber (b). (The author would like to thank Carissa Vetromile for this figure.)

Figure 4.4a shows the emission spectrum plotted in the most common format, namely intensity versus wavelength (excitation was at 470 nm using an ISS PC1 spectrofluorimeter). Such spectra can also be plotted as intensity versus wavenumber (or frequency) as shown in **Figure 4.4b**. Most modern spectrofluorimeters permit emission spectra to be displayed using wavelength or wavenumber scales. Wavenumber plots, though, are much less common in the literature than wavelength plots, even though they have the advantage of a linear energy scale on the X-axis (interestingly, wavenumber scales are commonly used in other branches of spectroscopy, for example, infrared absorption spectroscopy). If a photon-counting instrument is used to obtain the spectrum, one often notes that the Y-axis is labeled counts per second or cps. Sometimes, if an analog instrument is utilized, the Y-axis is labeled as "voltage," although this convention is rare. Quite often, the Y-axis is given in "arbitrary units." Why are emission (and excitation) intensities reported in this manner? Absorption values have specific units, namely *absorbance* or *optical density* (as defined in Chapter 2), which are independent of instrument parameters such as lamp intensity, slitwidths, PMT voltages and amplifier gains, since the measurements are always referred to the intensity of the incident light. In the case of fluorescence, however, the observed fluorescence intensity will depend upon how much light illuminates the sample—which is a function of the lamp power, the excitation slitwidth, and the efficiency of the excitation monochromator. The observed intensity will also depend on the emission slitwidth, the efficiency of the emission monochromator, the PMT voltage, the details of the electronic amplification (e.g., analog versus photon counting) and the particular optical arrangement of the instrument, that is, the focal length of the lenses and even the nature of the cuvette utilized, that is, pathlength. These parameters vary from instrument to instrument, and sometimes even from measurement to measurement. Thus, even though the signal being recorded may actually be the number of photons detected, these counts will vary with extrinsic instrumentation parameters in addition to the fluorophore concentration and photophysical parameters (e.g., extinction coefficient and quantum yield) and hence the descriptor "arbitrary" is utilized. It is also common to see the fluorescence intensity normalized to unity, that is, the maximum is set to "1."

Figure 4.5 shows emission spectra for fluorescein excited at three different wavelengths, namely, 280, 340, and 470 nm. The three emission spectra are presented as they appear after simply changing the excitation wavelength, while keeping all other parameters constant (left spectrum), and also normalized to the same maximum intensity (right spectrum). The difference in fluorescence intensities, before normalization, is due to the fact that the fluorescein absorbs more light at 470 nm than it does at 340 nm or 280 nm but also due to the fact that the light from the xenon arc source increases in the order 470 > 340 > 280 nm (see Figure 3.4). The normalized spectra illustrate the fact that the shape of the emission spectrum and its maximum are independent of the excitation wavelength. If the emission maximum actually changes with excitation wavelength it usually means that the fluorescent sample is not pure, that is, more than one fluorescing molecule is present which has a different absorption spectrum and/or emission spectrum than the target fluorophore.

These general observations, that is, independence of emission from the excitation wavelength, the shift between the excitation and emission maxima (known as the Stokes shift), as well as the approximate mirror-image aspect mentioned above, follow from consideration of the Perrin–Jabłoński diagram shown in Figure 4.6. I should note here that it is common to see this diagram referred to simply as the Jabłoński diagram. I prefer the nomenclature used by Bernard Valeur in his excellent treatise, *Molecular Fluorescence*, namely, the Perrin–Jabłoński diagram, to credit the contribution of Francis Perrin, who actually made energy diagrams before, and independent of, Alexander Jabłoński.

We recall the Franck–Condon principle from Chapter 2, namely, that the excitation process occurs so rapidly (on the order of 10^{-15} s), that no rearrangement of the nuclear framework occurs during this process. Although the fluorophore may be excited into different singlet state energy levels

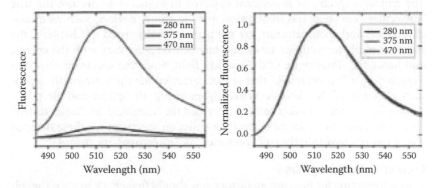

FIGURE 4.5 Comparison of fluorescein emission spectra excited at 280, 375, and 470 nm. Left: not normalized. Right: normalized.

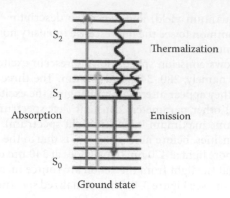

S_2

Thermalization

S_1

Absorption   Emission

S_0

Ground state

FIGURE 4.6 Perrin–Jabłoński diagram.

(e.g., S_1, S_2, etc.), rapid thermalization (on the order of 10^{-12} s) invariably occurs and emission takes place from the lowest vibrational level of the first excited electronic state (S_1). This fact accounts for the independence of the emission spectrum from the excitation wavelength. The fact that ground state fluorophores, at room temperature, are predominantly in the lowest vibrational level of the ground electronic state (as required from Boltzmann's distribution law; Equation 2.7) accounts for the Stokes shift. Finally, the fact that the spacings of the energy levels in the vibrational manifolds of the ground state and first excited electronic state are usually similar accounts for the fact that the absorption spectrum and the emission spectrum (plotted in energy units such as reciprocal wavenumbers) are approximately mirror images. This Perrin–Jabłoński diagram does not, however, indicate the excited state processes that may occur in some cases, such as dipolar relaxation, intersystem crossing, or FRET. Some of these processes will be covered in later chapters.

Correcting Emission Spectra

The emission spectra of fluorescein depicted in **Figure 4.4a** are not the true "molecular" emission spectra. Optical and electronic devices, such as monochromators and photodetectors, are not perfect. As discussed in Chapter 3, the response of photomultiplier tubes and monochromators vary with the energy, and hence the wavelength, of the incident light. Since the spectrum shown in **Figure 4.4a** is "uncorrected," that is, not corrected for the wavelength-dependent response of the detector system (comprising all optical and electronic components in the emission path), it is termed the "technical" or "uncorrected" spectrum. How, then, do we correct the emission spectrum for these detector characteristics to obtain the true "corrected" or "molecular" spectrum?

Correcting for Background

Before correcting for instrument factors, one should first check to see if the solvent itself contributes significantly to the fluorescence signal. If the fluorescence is much stronger than the signal due to the background, which may originate from contaminants in the solvent (usually a buffer) and/or from scatter peaks (mainly Raman scatter to be discussed shortly), then background subtraction may be unnecessary. The background must be taken with exactly the same instrument settings, for example, monochromator slitwidths, PMT voltage, amplifier gain, used for the sample run. On photon counting instruments one cannot alter the PMT voltage or amplifier gains, but on analog instruments these electronic parameters must be taken into account. Of course, this discussion implicitly assumes that the spectrofluorimeter is interfaced with a computer and has software to enable mathematical manipulation of the data. Some instruments include automatic background subtraction routines in their software. For example, such routines typically direct the user to place a cuvette containing the background in one cuvette holder and the sample in another. As the measurements are being obtained, the instrument can cycle between the sample and background and automatically display the sample signal minus the background. Although such routines can in principle be useful, I caution the novice not to depend too strongly on such automatic corrections. One problem is that one assumes that the two cuvettes are perfectly matched. But, also, I am a firm believer in observing directly the nature and magnitude of the background so that one can judge how severe the correction actually is, for example, is the background 5% of the signal or 85%? If the background is truly significant, one is better off working at improving the signal-to-background ratio.

Raman Peaks

Water molecules exhibit two O–H stretching frequencies (symmetric and asymmetric) and one bending frequency. These motions result in absorption of electromagnetic energy in the infrared region of the spectrum over a narrow range near 3400 cm^{-1}. Although the absorbance of these frequencies is weak, given the fact that water is so concentrated (55 M), the total absorption is not negligible, which results in a slight loss of energy in the scattered light from an aqueous solution. This energy loss results in an "inelastic scattering," known as Raman scatter (after Chandrasekhara Venkata Raman), shifted around 3400 cm^{-1} to lower energy from the elastic scattering known as Rayleigh scatter (after John William Strutt who, upon his father's death, became the 3rd Baron Rayleigh). A simple expression to estimate the position of the Raman scatter from water is

$$\frac{1}{\lambda_r} = \frac{1}{\lambda_e} - 0.00034 \tag{4.1}$$

where λ_r is the wavelength of the Raman scatter and λ_e is the wavelength of the excitation. **Figure 4.7** shows the position of the Raman peak for water, as a function of the excitation wavelength.

Sometimes, if one is concerned that a Raman peak is present in a spectrum, one can slightly alter the excitation wavelength and then a Raman peak will move whereas the fluorescence will remain constant. This approach is illustrated in **Figure 4.8**, which shows a dilute solution of tryptophan excited at 270, 280, and 290 nm. The resulting spectra are normalized so that one can

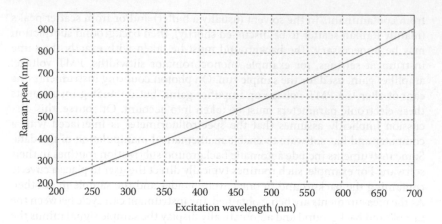

FIGURE 4.7 Plot of position of Raman peak of water versus excitation wavelength.

clearly see that the position of the tryptophan emission is invariant, while the Raman peaks move depending on the excitation.

Although the Raman peak can lead to confusion in fluorescence studies, it can provide some service as an emission standard. Specifically, the strength of the water Raman scatter can serve as a measure of the light intensity from a light source, for example, a xenon arc lamp, to indicate when the lamp is delivering less light than expected. For example, a log book can be kept near the spectrofluorimeter and every week or so one can monitor the Raman scatter intensity, near 397 nm, from a water solution excited at 350 nm, under the same conditions of slitwidths and lamp current settings.

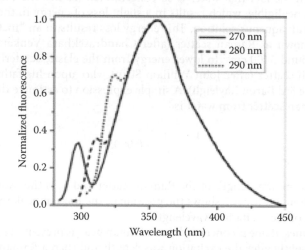

FIGURE 4.8 Uncorrected emission spectra of BSA with Raman peaks; excitation wavelengths were 270 nm (solid line), 280 nm (dashed line), and 290 nm (dotted line).

Correcting for Instrumental Parameters

Let us assume now that our technical emission spectrum has been corrected for any background contribution. What's next? When I was a graduate student in Gregorio Weber's lab, I had to generate corrected spectra on the new, photon-counting instrument we had constructed. To accomplish this task, I bought a "standard lamp" which had been calibrated against a National Bureau of Standards secondary source. When this lamp was operated at a precise voltage and current, the "color temperature" of the lamp was known. I could then use it to generate a "lamp curve" and compare these data to those supplied by the NBS. In this way, correction factors for the detector/monochromator system of our instrument could be determined. Nowadays, such correction factors are usually supplied with the spectrofluorimeter, and so one only has to use the software to apply the corrections and hence obtain the true molecular spectrum of the sample. Of course, if for any reason the emission PMT needs to be replaced, then, in principle, new corrections should be generated. Another approach now available is to utilize known standards supplied with predetermined corrected or molecular spectra. These are available from the German Bundensanstalt für Materialforschund und -prüfung (BAM; Federal Institute for Materials Research and Testing) and can be purchased from Sigma-Aldrich.

Let us look at an uncorrected and corrected emission spectrum. **Figure 4.9a** shows the uncorrected or technical emission for ANS in ethanol, obtained on an ISS PC1 spectrofluorimeter. The first step toward the corrected emission spectrum is to obtain the emission spectrum with a polarizer oriented vertically (i.e., parallel to the vertical laboratory axis) inserted in the emission light path. The small bump in the technical spectrum, near 507 nm, is the Wood's anomaly mentioned in Chapter 3 which, as the reader no doubt recalls, means that at this wavelength some of the horizontal (perpendicular polarized) component of the emission is lost. So before we apply correction factors, it is convenient to remove the Wood's anomaly. Again, as discussed earlier, we simply view the emission through a polarizer oriented to pass vertically polarized (parallel) light. This emission spectrum, viewed through a parallel polarizer, is shown in **Figure 4.9b** (solid line). Now, we simply apply the correction factors supplied with the instrument for parallel polarized light (and for the slit-width utilized while recording the spectrum). These correction factors, for vertical (parallel) polarized emission, are shown in **Figure 4.9c**.

Occasionally, I am asked "Do I always have to correct an emission spectrum?" In fact, I advise people not to worry about correcting spectra in their day-to-day experiments. For almost all purposes, technical or uncorrected spectra will suffice—it is only when one is about to publish a spectrum that one really needs to think about corrections. In fact, as a frequent reviewer of manuscripts, I don't mind if authors submit an uncorrected spectrum as long as they state that it is an uncorrected or technical spectrum and they state what instrument they used. Given modern red-sensitive photodetectors, the difference between an uncorrected (or technical) emission spectrum and the corrected (or molecular) emission spectrum may, in fact, not be very large. Of course, if they are using the spectrum to calculate some other parameter, such

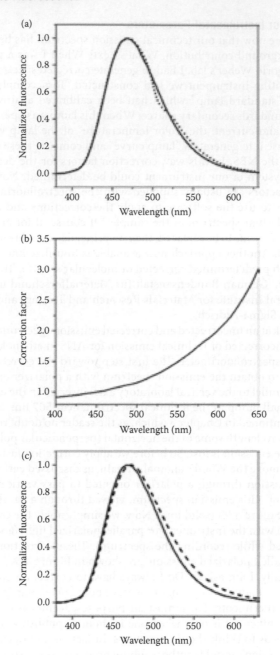

FIGURE 4.9 Uncorrected and corrected emission spectra of ANS in ethanol. (a) Uncorrected emission spectrum (dotted line) and uncorrected emission spectrum viewed through parallel polarizer (solid line). (b) Correction factors for parallel polarizer. (c) Uncorrected emission spectrum viewed through parallel polarizer (solid line) and corrected emission spectrum (dashed line).

as the quantum yield or Förster overlap integral (more on these topics later), that is another story, and in such cases corrected spectra must be utilized. I have also been asked "Since my instrument allows me to generate a corrected spectrum in real time, i.e., while the data is being recorded, why shouldn't I just always operate in that mode?" As I mentioned earlier in the discussion about background subtraction, I think it is always preferable to view the raw, that is, uncorrected, data so that one can judge the nature and magnitude of the corrections being applied.

Spectral Center of Mass

In some cases, changes in the position of the emission maxima, which accompany some process such as ligand binding, oligomerization, or denaturation, are of great interest. The most common way to determine an emission maximum is to look at the recorded spectrum and estimate the wavelength corresponding to the highest signal, that is, estimate λ_{max}. Such "eyeball" estimates may be adequate for many purposes, especially if the wavelength shifts are significant. In some cases, however, one may desire a less subjective and more sensitive method. In such cases, one may consider determining the spectral center of mass, $\langle v_g \rangle$, defined by the relation:

$$ v_g = \frac{\sum v_i F_i}{\sum F_i} \tag{4.2} $$

where F_i is the emission at a wavenumber, v_i, and the sum is carried over all wavenumbers where $F_i > 0$. This approach offers a very precise and reproducible measure of spectral shifts. For example, using Equation 4.2, the calculated center of mass for the spectrum shown on the right-hand side of **Figure 4.4** is 19, 277 cm^{-1}. Apparently, the term "center of mass" dates back to the days of spectroscopy before computer interfaced equipment, when spectra were recorded on chart paper, and people would cut out the spectrum and weigh the paper to get the area, and then cut the spectrum into parts and weigh the pieces to find the "center of mass." An example of the use of this approach is in the article by Silva et al. (1986) (*Biochemistry* 25:5780–5786), which followed the dissociation of the tryptophan synthase dimer induced by elevated hydrostatic pressure. The dissociation into monomers leads to a red shift in the intrinsic protein fluorescence, as shown in **Figure 4.10a**, which can be quantified using the center of mass. A plot of the center of mass versus applied pressure (**Figure 4.10b**), for example, allows one to readily follow the dimer dissociation process.

Excitation Spectra

An excitation spectrum measures the relative efficiencies of different incident wavelengths to elicit fluorescence. As stated earlier, the emission spectrum of a fluorophore is independent of the excitation wavelength. This fact was

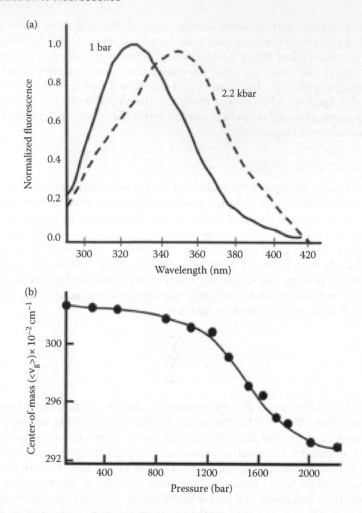

FIGURE 4.10 (a) Emission spectra of the holo form of tryptophan synthase at pressures of 1 bar and 2.2 kbars. (b) Illustration of center-of-mass used for quantification of wavelength shifts. (Adapted with permission from J.L. Silva et al., 1986. *Biochemistry* 25: 5780. Copyright 1996 American Chemical Society.)

illustrated in **Figure 4.5**, which also demonstrates that the intensity of the emission varies depending on the excitation wavelength. In fact, for a pure, well-behaved fluorophore (and we shall indicate what this cryptic term "well-behaved" means as we go along) the excitation spectrum should correspond to the absorption spectrum. This fact follows since the greater the extinction coefficient of a fluorophore, at any given wavelength, the greater the probability that it will absorb light and reach the excited state. But the prescient reader may also note that this correspondence between absorption and excitation spectra only holds if the quantum yield of the fluorophores is independent of the exciting wavelength! This caveat usually holds true—but we shall discuss exceptions later in this book when we explicitly discuss quantum yields.

To record an excitation spectrum, the excitation monochromator is scanned while the emission observed is kept constant, either by keeping the emission monochromator fixed or by viewing the emission through an appropriate filter (intereference, bandpass, or longpass). In this way, the relative efficiencies of the different wavelengths of incident radiation to excite the fluorescence may be determined. In the previous section on emission spectra, I stated that the difference between an uncorrected and a corrected emission spectrum may not be very large. The difference between an uncorrected and a corrected excitation spectrum, however, can be considerable!

To illustrate this point, **Figure 4.11a** shows the absorption spectrum of ANS in ethanol along with the uncorrected excitation spectrum. The uncorrected excitation spectrum was obtained on an ISS PC1 spectrofluorimeter (ISS, Inc., Champaign, IL) but the result would be similar on most instruments, be they commercial or homebuilt. Clearly ANS absorbs more strongly at 270 nm than it does at 360 nm, yet in the uncorrected excitation spectrum, we clearly see that illumination at 270 nm does not give rise to as much fluorescence as does 360 nm illumination. Why? The answer lies in the nature of the light coming out of the xenon arc lamp. As shown in Chapter 3 in Figure 3.4, the output of this lamp varies dramatically with wavelength. It is thus immediately apparent that the intensity reaching the sample upon 270 nm excitation is much less than at 360 nm excitation. Hence, if we wish to construct a "true" excitation spectrum we must take into account the variation of the lamp intensity with wavelength. I should note at this point that monochromators do not transmit all wavelengths with equal efficiencies and that the lamp spectrum shown above is actually recorded after passing through the excitation monochromator. But the major effect, by far, is the intrinsic variation of the xenon arc output with wavelength.

How can we carry out the necessary correction? An initial correction can be realized by using a device to directly monitor the intensity coming through the lamp/excitation monochromator as the wavelength is varied. One could then, in principle, use this signal to correct the excitation scan, for example, by recording the ratio of the sample signal divided by the signal from the device monitoring the exciting intensity. Such a correction is shown in **Figure 4.11a**.

How does one obtain this reference signal? Although you may suppose that one simply sets up a photomultiplier to observe some of the excitation light using a beam-splitter to deflect some of the main excitation beam toward this reference PMT, this approach will not work. The problem, as already indicated in the previous chapter, is that PMT's have a wavelength-dependent response and hence the PMT output will not faithfully record the variation in photon density with excitation wavelength but will convolute that signal with the PMT's wavelength-dependent efficiency. To circumvent this difficulty one must use a "quantum counter," which is to say a device which can monitor all incident wavelengths with equal efficiencies. In the particular correction shown in **Figure 4.10b**, the device monitoring the excitation was, in fact, a so-called "quantum counter." The principle underlying the quantum counter is illustrated in **Figure 4.12**.

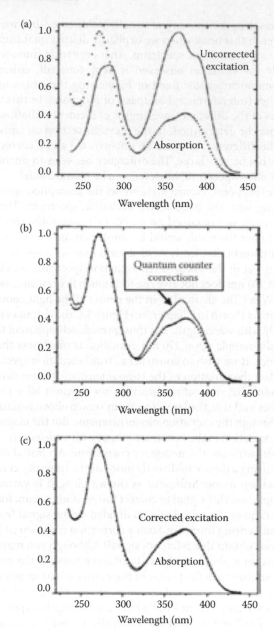

FIGURE 4.11 (a) Absorption spectrum (dotted line) and uncorrected excitation spectrum (solid line) for ANS in ethanol. (b) Absorption spectrum and quasi-corrected excitation spectrum (solid line) for ANS in ethanol. (c) Absorption spectrum (dotted line) and corrected excitation spectrum (solid line) for ANS in ethanol. (Adapted from D.M. Jameson, J.C. Croney, and P.D. Moens, 2003. *Methods Enzymol.* 360: 1–43.)

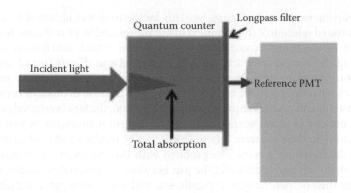

FIGURE 4.12   Schematic diagram of a quantum counter.

Basically, one takes a very concentrated solution of a high quantum yield fluorophore—rhodamine dyes are perhaps the most common ones used. The concentration must be high enough to insure that all photons at all wavelengths utilized will be absorbed by the sample. Note that the Stokes shift insures that even if all the incident light, and some of the emission, is absorbed, there will still be emission that is not absorbed. In practice, one may use around 3–5 mg of rhodamine per milliliter of ethanol or ethylene glycol (aqueous solutions are not used since the xanthene dyes have a propensity to form aggregates in water). The emission is observed through a filter so that the "redder" part of the spectrum is observed (although given the high concentration of the fluorophores, the emission will already be red-shifted by the inner filter effect). The signal from this reference channel should thus faithfully report on the variation of the photon flux from the excitation source as a function of wavelength. By dividing the sample signal by this reference signal, one should thus "correct" the recorded excitation spectrum for variations of the lamp intensity with wavelength. As shown in **Figure 4.11b**, this method works fairly well, although (in the example given) not perfectly.

So, you may ask, why didn't this correction work perfectly? The problem is that this "real-time" correction implicitly assumes that the beamsplitting device is ideal, that is, that the fraction of the excitation light deflected towards the reference channel is constant, regardless of the wavelength. In fact, perfection is rarely achieved—even in Introduction to Fluorescence texts! As indicated in the previous chapter, the polarization of the light exiting a monochromator varies with wavelength, and, hence, the polarization of the light hitting the beamsplitter varies with wavelength. But the parallel and perpendicular polarized components are not reflected with identical efficiencies. This difference is particularly noticeable when the wavelength corresponding to the Wood's anomaly of the excitation monochromator is reached—at which point most of the incident light is polarized parallel to the laboratory axis.

Now consider the corrected excitation spectrum for ANS in ethanol shown in **Figure 4.11c**. One notes the agreement between the absorption spectrum and the corrected excitation spectrum is quite good. How was this spectrum obtained? In this case, the uncorrected spectrum (shown in **Figure 4.11a**) was divided by the lamp curve shown in Figure 3.4. This lamp curve was obtained by using a concentrated solution of rhodamine B in ethanol placed in a small cuvette (3 mm square) in the sample compartment. Hence, the actual distribution of light reaching the sample compartment was measured and thus is free of any artifacts introduced by using the beamsplitter. Of course, this procedure requires that two separate spectra be obtained—but given the ease of modern computerized instruments and subsequent data manipulation, the task is relatively simple. I have noticed in recent years that some instrument manufacturers will store a lamp curve in their software files so that it can be recalled and used to generate corrected excitation spectra. The problem with this approach is that xenon arc lamps definitely age, specifically the gap between the electrodes widens, and as they age their output changes. Typically, one will note less output in the ultraviolet region of the spectrum as the lamp ages and hence a "stored" lamp curve may not accurately reflect the current energy output if many months have passed.

One may now ask "Why should we want to determine an excitation spectrum if it matched the absorption spectrum?" In fact, there are several reasons why an excitation spectrum may be useful. First of all, if one has a mixture of chromophores, only one of which emits fluorescence, one can utilize the excitation spectrum to obtain the absorption spectrum of the chromophore even in a mixture of other absorbing molecules. A more interesting situation occurs, however, when the excitation spectrum does not, in fact, match the absorption spectrum. This situation could signify that the fluorophore solution is not pure or that an excited state process, such as FRET, is occurring. One common example is protein fluorescence. The absorption spectrum of a protein is largely due to tyrosine and tryptophan residues. However, if one monitors the tryptophan fluorescence, for example, at 340 nm, one will usually note that the corrected excitation spectrum does not match the absorption spectrum of the protein. The reason for this difference is that tyrosine to tryptophan energy transfer can occur and hence some of the light absorbed by the tyrosine residues can lead to tryptophan excitation. This topic will be discussed in more detail in Chapter 11.

Delayed Fluorescence

In rare cases, an excited molecule can exist in a triplet state (which is the state that gives rise to phosphorescence), but revert back to the excited singlet state and then emit fluorescence. This emission is known as "delayed fluorescence" since it may occur at times very long after normal fluorescence, even on the microsecond to millisecond time scale. Since this phenomenon is very unusual, we shall not discuss it in any detail.

Advanced Scanning Methods

Finally, I wish to mention two other types of steady-state scanning methods used primarily, though not exclusively, in analytical chemistry. These are (1) excitation–emission matrix (EEM) approaches and (2) synchronous scanning methods.

EEM

This approach, introduced by Gregorio Weber in 1961, entails acquisition of emission spectra at varying excitation wavelengths. Alternatively, one can think of the acquisition of excitation spectra at varying emission wavelengths. The rationale, as first delineated by Weber more than 50 years ago, is to utilize the fact that different fluorophores will invariably exhibit either differing excitation or emission spectra or both. Hence, a matrix of intensity values obtained over a range of both exciting and emitting wavelengths will be unique for each specific mixture of fluorophores. In Weber's early work, the motivation was to ascertain how many components were present in a mixture. In many cases, such data can be used to identify the components of the mixtures. Typically, in analytical applications, EEMs are used to obtain a unique "fingerprint" of a mixture, which can be used for sample characterization or even forensic purposes. The EEM method has, in fact, found wide application in environmental issues such as organic matter in water (including waste water), soil, and oils. An example of an EEM is shown in **Figure 4.13**. In this example, ANS is bound to BSA and one notes three peaks due to BSA, ANS excited directly by mid-UV, ANS excited near 280 nm as well as ANS excited via energy transfer from BSA.

Synchronous Scanning

This method is also used to examine mixtures of fluorophores but the approach is quite different from the EEM method. In synchronous scanning, both excitation and emission monochromators are scanned simultaneously with a

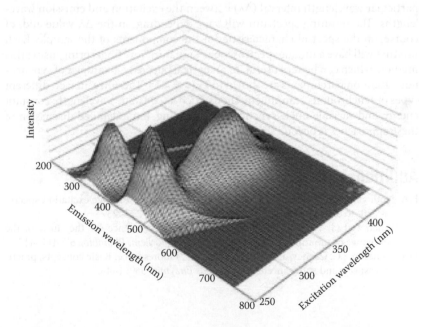

FIGURE 4.13 Excitation emission matrix (EEM) of ANS bound to BSA.

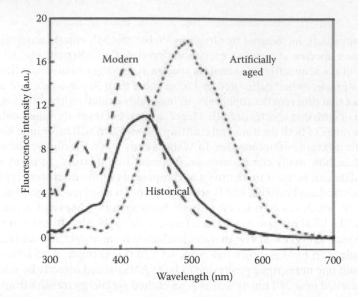

FIGURE 4.14 Synchronous scans of modern, historical, and artificially aged parchment samples. The wavelength difference between excitation and emission was 70 nm. (From B. Dolgin, V. Bulatov, and I. Schechter, 2009. *Anal. Bioanal. Chem.* 395: 2151.)

particular wavelength interval ($\Delta\lambda$) between the excitation and emission wavelengths. The resulting spectrum will vary depending on the $\Delta\lambda$ value and, of course, on the spectral characteristics of the components of the sample. Each mixture will have a unique spectroscopic "signature" or "fingerprint" using this method, which can be very useful for analytical purposes. Examples of synchronous scans on parchments are shown in **Figure 4.14**. There are, in fact, different types of synchronous scanning methods, such as constant wavelength, constant energy, and variable-angle synchronous scanning, but a detailed discussion of these approaches is beyond the scope of this book.

Additional Reading

E.A. Nothnagel, 1987. Quantum counter for correcting fluorescence excitation spectra at 320- to 800-nm wavelengths. *Anal. Biochem.* 163: 224–237.

L. Erijman and G. Weber, 1993. Use of sensitized fluorescence for the study of the exchange of subunits in protein aggregates. *Photochem. Photobiol.* 57: 411–415.

D.M. Jameson, J.C. Croney, and P.D. Moens, 2003. Fluorescence: Basic concepts, practical aspects and some anecdotes. *Methods Enzymol.* 360: 1–43.

5

Polarization and Anisotropy

AS MENTIONED IN CHAPTER 3, Etienne-Louis Malus coined the term "polarization" to describe a specific property of light. George G. Stokes was probably the first to carry out a fluorescence polarization measurement. He isolated the emission of quinine from the exciting light using colored filters, and viewed this emission through Nicol polarizers. He reported that there was no apparent polarization, that is, no change in intensity upon rotation of the polarizer, which was essentially correct since the lifetime of quinine is ~20 ns, sufficiently long such that reorientation of the excited fluorophore is virtually complete during its excited state lifetime (the relationship between polarization and lifetime will be discussed in detail later in this chapter). A careful measurement using modern instrumentation would find a value of ~0.001 for the polarization of quinine—a value too small to be discerned by the visual comparison method available to Stokes. Interestingly, had Stokes observed the polarization of the fluorescence from *Lignum nephriticum* he may well have found a noticeable polarization since the lifetime of the fluorophore (~2.8 ns) is much shorter than that of quinine and would have rendered a polarization in the range of 0.07.

In 1920, F. Weigert discovered that the fluorescence from solutions of certain dyes was polarized. Specifically, he looked at solutions of fluorescein, eosin, rhodamine, and other dyes, and noted the effect of temperature and viscosity on the observed polarization. In Weigert's words "Der Polarisationsgrad des Fluorezenzlichtes nimmt mit wachsender Molekulargröße, mit zunehmender Viskosität des Mediums und mit abnehmender Temperatur, also mit Verringerung der Beweglichkeit der Einzelteilchen zu." *"The degree of the polarization increases with increasing molecular size, with increasing viscosity of the medium and with decreasing temperature, that is with the reduction of the mobility of the single particles."* He recognized that all of these considerations meant that fluorescence polarization increased as the mobility of the emitting species decreased.

The first comprehensive study of this newly discovered phenomenon was due to S. I. Vavilov and V. L. Levschin in 1923, who measured the polarization of 26 dyes in water and glycerol. They were able to demonstrate that some of the dyes they studied showed large differences between polarizations in water compared to the polarizations in glycerol, whereas other dyes gave similar polarizations regardless of the solvent's viscosity (we will learn later in this chapter that these facts depend upon the lifetime of the dyes examined). These prescient observers, in fact, postulated that the fluorescence was due to molecular rotation of a fluorophore and was characterized by an electric vector that could oscillate only in one direction, which led them to correctly calculate that the maximum values of the polarization would be 1/2 for such a linear oscillator and 1/7 for a circular oscillator (I shall discuss this so-called "limiting" polarization later in this chapter).

We should note that a very important paper in the history of polarization and energy transfer appeared in 1924—by Enrique Gaviola and Peter Pringsheim. Specifically, they showed that the polarization of sodium fluorescein in glycerol (which was very viscous so as to hinder rotation of the fluorescein molecules) was essentially zero at very high concentrations of fluorescein (they actually managed to dissolve 160 mg of sodium fluorescein in 1 g of glycerol!) but became progressively higher as the fluorescein concentration was decreased—finally reaching a value of 0.45 when the starting solution was diluted about 20,000 fold.

In 1925–1926, Francis Perrin published several important papers describing a quantitative theory of fluorescence polarization, including a classic paper containing most of the essential information that we use to this day. Polarization remained largely in the province of the physicists for almost two decades, until Gregorio Weber began his thesis work with the famous enzymologist Malcolm Dixon in Cambridge in the mid-1940s. Weber's subsequent theoretical and experimental work—which extended Perrin's earlier contributions and also developed what became modern instrumentation—brought fluorescence polarization to the attention of the biochemistry community, and so ushered in a new scientific discipline—quantitative biological fluorescence. Gregorio Weber once wrote: "It was from reading Perrin's papers that I conceived three ideas on the use of polarization: determination of the change in the fluorescence lifetime as one quenches the fluorescence by addition of an appropriate chemical, determination of the molecular volume of proteins by fluorescent conjugates with known dyes and determination of the viscosity of a medium through the polarization of the emission from a known fluorescent probe."

Weber began to apply polarization to biochemistry, first by looking at the polarization of flavin compounds, both free and associated with proteins, and then by attaching a fluorescent probe, dansyl chloride (which he had to synthesize), to proteins in order to study the proteins' hydrodynamic properties. At the same time, in the early 1950s, his colleague, David J.R. Laurence, using Weber's instrumentation, studied the binding of fluorescent molecules such as fluorescein, eosin, 5-amonoacridine, and other dyes, to bovine serum albumin. Laurence thus developed the homogeneous fluorescence binding assay; that is, an assay which did not require separation of free and bound ligand (more about this type of assay later).

Basic Principles

As stated in Chapter 2, light can be considered as oscillations of an electromagnetic field, characterized by electric and magnetic components, perpendicular to the direction of light propagation. We shall be concerned only with the electric component. In natural light, the electric field vector can assume any direction of oscillation perpendicular or normal to the light propagation direction, as shown in **Figure 5.1** (top). Polarizers are optically active devices that can isolate one direction of the electric vector, as illustrated in **Figure 5.2** (bottom).

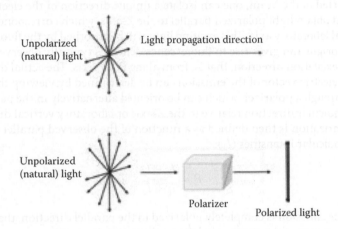

FIGURE 5.1 Depiction of unpolarized or natural light, before and after passing through a polarizer.

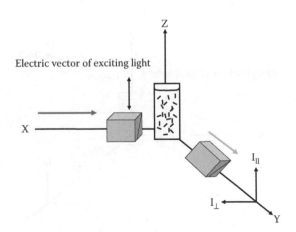

FIGURE 5.2 Depiction of polarized light impinging upon a sample and observation of the fluorescence from the sample through polarizers oriented in the vertical and horizontal directions relative to the laboratory axes.

The most common polarizers used today are (1) dichroic devices, which operate by effectively absorbing one plane of polarization (e.g., Polaroid type-H sheets based on stretched polyvinyl alcohol impregnated with iodine) and (2) double refracting calcite ($CaCO_3$) crystal polarizers—which differentially disperse the two planes of polarization (examples of this class of polarizers are Nicol polarizers, Wollaston prisms, and Glan-type polarizers such as the Glan–Foucault, Glan–Thompson, and Glan–Taylor polarizers).

Consider an XYZ coordinate framework with a fluorescent solution placed at the origin, as shown in **Figure 5.3**, where XZ is in the plane of the page. In this system, the exciting light is traveling along the X direction. If a polarizer is inserted in the beam, one can isolate a unique direction of the electric vector and obtain light polarized parallel to the Z-axis, which corresponds to the vertical laboratory axis. This exciting light will be absorbed by the fluorophore at the origin and gives rise to fluorescence, which is typically observed at 90° to the excitation direction, that is, from along the Y-axis. The actual direction of the electric vector of the emission can be determined by viewing the emission through a polarizer, which can be oriented alternatively in the parallel or perpendicular direction relative to the Z-axis or laboratory vertical direction.

Polarization is then defined as a function of the observed parallel ($I_{\parallel}$) and perpendicular intensities ($I_{\perp}$):

$$P = \frac{I_{\parallel} - I_{\perp}}{I_{\parallel} + I_{\perp}} \tag{5.1}$$

If the emission is completely polarized in the parallel direction, that is, the electric vector of the exciting light is totally maintained, then

$$P = \frac{1 - 0}{1 + 0} = 1 \tag{5.2}$$

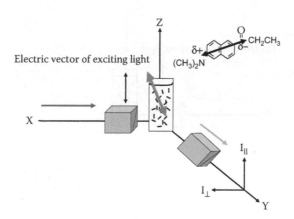

FIGURE 5.3 Depiction of a transition dipole moment on a nuclear framework. The large dipole in the solution depicts a dipole oriented relative to the polarization direction of the excitation.

If the emitted light is totally polarized in the perpendicular direction, then

$$P = \frac{0 - 1}{0 + 1} = -1 \tag{5.3}$$

The limits of polarization in these cases would be +1 to −1.

Another term frequently used in the context of polarized emission is anisotropy (usually designated as either r or A) which is defined as

$$r = \frac{I_{\parallel} - I_{\perp}}{I_{\parallel} + 2I_{\perp}} \tag{5.4}$$

By analogy to polarization, the mathematical limits of anisotropy are +1 to −0.5.

Given the definition of polarization and anisotropy, one can show that

$$r = \frac{2}{3}\left(\frac{1}{P} - \frac{1}{3}\right)^{-1} \quad \text{or} \quad r = \frac{2P}{3 - P} \tag{5.5}$$

For example:

P	r
0.50	0.40
0.25	0.20
0.10	0.069
−0.10	−0.065

Clearly, the information content of the polarization function and the anisotropy function are essentially identical and the use of one term or the other is dictated by practical considerations, which will be discussed later.

In solution, these limits (e.g., ±1 for polarization) are not realized. The reasons underlying this fact have to do with the principle of photoselection as well as the distribution of orientations for a population of fluorophores. Consider, as shown in **Figure 5.3**, fluorophores at the origin of our coordinate system.

Upon absorption of an exciting photon, the electron cloud associated with the molecule is altered and a dipole moment is created, as indicated in the figure (usually of different magnitude and direction from the ground state dipole). The orientation of this dipole moment, relative to the nuclear framework, and its magnitude, will be determined by the nature of the substituents on the molecule. This excited state dipole moment is also known as the transition dipole or transition moment. In fact, if light of a particular electric vector orientation (plane polarized light) impinges on a sample, only those molecules properly oriented relative to this electric vector can absorb the light. Specifically, the probability of the absorption is proportional to the cosine squared ($\cos^2\theta$) of the angle θ between the exciting light and the transition dipole (**Figure 5.4**).

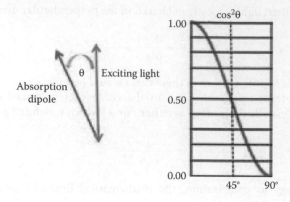

FIGURE 5.4 Left: Depiction of an absorption transition dipole oriented at an angle, θ, relative to the polarization direction of the excitation. Right: Depiction of the value of the cos²θ function over a range of 0 to 90 degrees.

Hence, when we excite an ensemble of randomly oriented fluorophores with plane-polarized light, we are performing a *photoselection* process, that is, creating a population of excited molecules which nominally have their excited dipoles lined up with the polarization direction of the excitation. This process is illustrated in **Figure 5.5**. Consider now that the transition dipole corresponding to the emission of light from the excited fluorophore is *parallel* to the absorption dipole (as indicated in **Figure 5.6**) and that the excited fluorophore cannot rotate during the lifetime of the excited state (e.g., if the fluorophores are embedded in a highly viscous or frozen medium). If we were now to measure the polarization of the emission it would be less than +1 since some of the dipoles excited will not be exactly

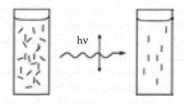

FIGURE 5.5 Depiction of the photoselection process.

FIGURE 5.6 Structure of a fluorophore depicting the parallel orientation of the absorption and emission dipole.

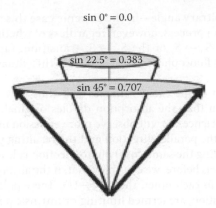

FIGURE 5.7 Depiction of the value of sin θ at different angles: thicker lines indicate more dipoles at that orientation.

parallel to the direction of the exciting light. In fact, the number of potential dipoles making an angle θ with the vertical axis will be proportional to sin θ (**Figure 5.7**).

We can then calculate that the upper polarization limit for such a randomly oriented (but rigidly fixed, i.e., nonrotating) ensemble—with co-linear excitation and emission dipole—will be +1/2. This calculation, however, is a bit beyond the level of this introductory text.

Consider the case in **Figure 5.8**. Here, the two principal absorption bands for phenol are depicted along with the emission band. The energy level diagram corresponding to this system is also depicted. The directions of the absorption dipoles—relative to the nuclear framework—may differ greatly for the two transitions as illustrated below. Hence, the two excited dipoles corresponding to the $S_0 \rightarrow S_1$ and the $S_0 \rightarrow S_2$ transitions may be

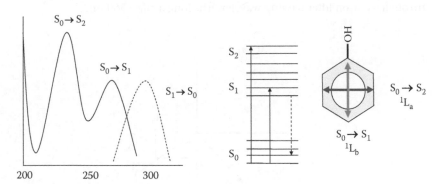

FIGURE 5.8 Left: Absorption and emission spectra for phenol. Middle: schematic of the ground and first two excited states of phenol. Right: sketch of phenol depicting the orientation of the transition dipoles corresponding to the $S_0 - S_1$ and $S_0 - S_2$ absorptions.

oriented at an arbitrary angle—in the extreme case this angle could be 90°. After the excitation process, however, regardless of whether absorption corresponded to the $S_0 \rightarrow S_1$ or the $S_0 \rightarrow S_2$ transition, rapid thermalization leaves the excited fluorophore in the S_1 level. This situation is depicted in **Figure 5.9**.

The orientation of the excited dipoles will thus now possess a different average orientation than the absorption dipoles originally photoselected by the exciting light. Hence, we will observe more emission in the perpendicular direction than in the parallel direction and the resulting polarization will be negative. Considering the same $\cos^2\theta$ photoselection rule and the $\sin\theta$ population distribution as before, we can show that, if the absorption and emission dipoles are at 90° to each other, then $P = -1/3$. These polarization values, in the absence of rotation, are termed limiting or intrinsic polarizations and are denoted as P_0. In general,

$$\frac{1}{P_0} - \frac{1}{3} = \frac{5}{3}\left(\frac{2}{3\cos^2\phi - 1}\right) \tag{5.6}$$

where ϕ is the angle between absorption and emission dipoles. We can now understand that the limiting polarization of a fluorophore *will depend upon the excitation wavelength—a very important point!* Let me say that again: *a very important point!*

Consider the excitation polarization spectrum for phenol (in glycerol at –70°C) in **Figure 5.10**. This spectrum corresponds to the system shown in **Figure 5.8** and is from the original data published by Weber. In cases wherein there are multiple overlapping absorption bands at various angles, the excitation polarization spectrum can be complex, as shown for indole (**Figure 5.11**).

Figure 5.12 shows the excitation polarization spectra of rhodamine B embedded in a Lucite matrix at room temperature—data I obtained when I was a graduate student in Gregorio Weber's laboratory. Emission was viewed through a cut-on filter passing wavelengths longer than 560 nm.

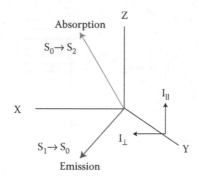

FIGURE 5.9 Representation of the orientations of the $S_0 - S_2$ absorption dipole and $S_1 - S_0$ emission dipole, relative to the vertical and horizontal emission polarizer orientations.

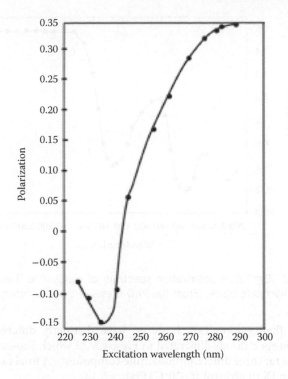

FIGURE 5.10 Excitation polarization spectrum of phenol in propylene glycol at −58°C. (Adapted from G. Weber, 1966. In *Fluorescence and Phosphorescence Analysis*, Ed. D. Hercules, Interscience Publishers, New York, pp. 217–240.)

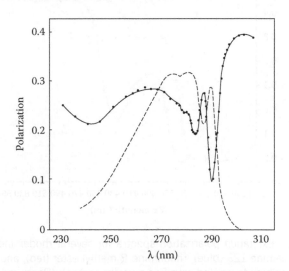

FIGURE 5.11 Excitation polarization spectrum of indole. (From D.M. Jameson et al., 1978. *Rev. Sci. Instru.* 49: 510.)

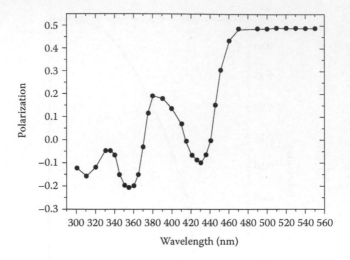

FIGURE 5.12 Excitation polarization spectrum of rhodamine B embedded in a methyl methacrylate block. (From the PhD thesis of D.M. Jameson.)

Similar fluorophores can, however, have distinctly different excitation polarization spectra—as illustrated in **Figure 5.13** which shows the polarization spectra for three different rhodamine compounds. A final example is protoporphyrin IX in glycerol at –20°C (**Figure 5.14**).

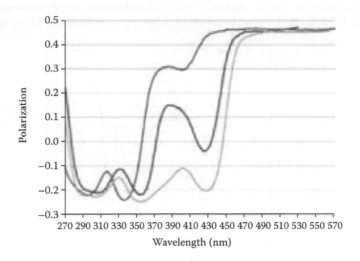

FIGURE 5.13 Excitation polarization spectra for several rhodamines in glycerol at 2°C. Rhodamine 123 (blue), rhodamine B methyl ester (red), and tetramethylrhodamine maleimide reacted with free cysteine (green). (Reprinted with permission from D.M. Jameson and J.A. Ross, 2010. *Chem. Rev.* 110: 2685–2708. Copyright 2010 American Chemical Society.)

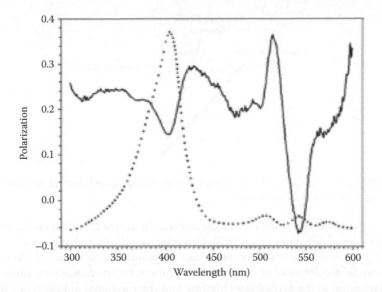

FIGURE 5.14 Excitation polarization spectrum of protoporphyrin IX in propylene glycol at minus 40°C. (Unpublished data of G.D. Reinhart and D.M. Jameson.)

Such polarization spectra are often obtained by viewing the emission through a long-pass filter that blocks the exciting light but passes most of the emission. This approach only works, of course, if the polarization is constant across the emission band, which is usually—though not always—the case. I must admit, though, that not everyone who contemplates using polarization is interested in solutions of fluorophores in glycerol at −50°C! We may thus want to consider the effects of fluorophore rotation on the observed polarization. We should note that the rotational diffusion of spherical molecules is described by the relation:

$$D = \frac{k_B T}{6\pi\eta r} \tag{5.7}$$

where D is the rotational diffusion coefficient, k_B is Boltzman's constant, T is the temperature, η is the solvent viscosity, and r is the radius of the molecule. Additional depolarization occurs if, during the excited state lifetime, the dipole rotates through an angle ω. In fact,

$$\frac{1}{P} - \frac{1}{3} = \left(\frac{1}{P_0} - \frac{1}{3}\right)\left(\frac{2}{3\cos^2\omega - 1}\right) \tag{5.8}$$

where P is the observed polarization. So the total depolarization is determined by an intrinsic factor (P_0) and an extrinsic factor (ω) (P. Soleillet first pointed out, in

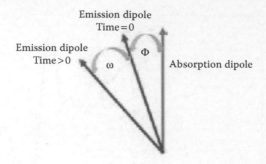

FIGURE 5.15 Depiction of the change in the relative orientation of an emission dipole with time after excitation.

1929, the properties of the factors [containing the cosine function] on the right-hand sides of Equations 5.6 and 5.8). This process is illustrated in **Figure 5.15**.

Figure 5.16 illustrates the processes of photoselection and rotation diffusion in a collection of fluorophores. Francis Perrin related the observed polarization to the excited state lifetime and the rotational diffusion of a fluorophore in his famous manuscript: Perrin, F. 1926. Polarisation de la Lumiere de Fluorescence. Vie Moyene des Molecules Fluorescentes. *J. Phys.* 7:390–401. Specifically,

$$\frac{1}{P} - \frac{1}{3} = \left(\frac{1}{P_0} - \frac{1}{3}\right)\left(1 + \frac{RT}{\eta V}\tau\right) \tag{5.9}$$

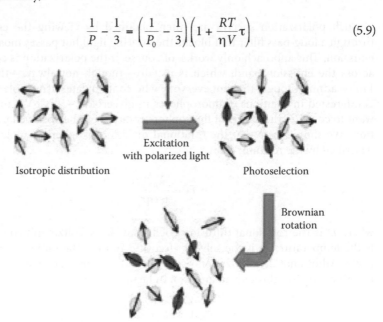

FIGURE 5.16 Illustration of the effect of photoselection and Brownian diffusion of a population of fluorophores. (Reprinted with permission from D.M. Jameson J.A. and Ross, 2010. *Chem. Rev.* 110: 2685–2708. Copyright 2010 American Chemical Society.)

where V is the molar volume of the rotating unit, R is the universal gas constant, T the absolute temperature, η the viscosity, and τ the excited state lifetime. We can rewrite this equation as

$$\frac{1}{P} - \frac{1}{3} = \left(\frac{1}{P_0} - \frac{1}{3}\right)\left(1 + \frac{3\tau}{\rho}\right) \tag{5.10}$$

where ρ is the Debye rotational relaxation time, which is the time for a given orientation to rotate through an angle given by the arc cos e^{-1} (68.42°). For a spherical molecule:

$$\rho = \frac{3\eta V}{RT} \tag{5.11}$$

For a spherical protein, it follows that:

$$\rho_0 = \frac{3\eta M(\upsilon + h)}{RT} \tag{5.12}$$

where M is the molecular mass, υ is the partial specific volume, and h the degree of hydration. *Hence, the larger the rotating unit, the larger the Debye rotational relaxation time and the larger the polarization!* This principle is illustrated in **Figure 5.17** for a fluorophore free in solution, compared to one bound to a protein.

I should note that it is not uncommon to see the term "rotational correlation time," often denoted as τ_c, used in place of ρ, the Debye rotational relaxation time. The information content of these terms are essentially identical

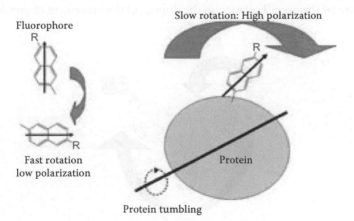

FIGURE 5.17 Illustration of the fast rotation, and hence low polarization, of fluorophores free in solution compared to the slow rotations, and hence high polarization, of fluorophores linked to a protein. (Reprinted with permission from D.M. Jameson and J.A. Ross, 2010. *Chem. Rev.* 110: 2685–2708. Copyright 2010 American Chemical Society.)

since $\rho = 3\tau_c$, but I have observed that some people become rather fervently attached to the use of one term or the other. In the original development of the theories of rotational motion of fluorophores, Perrin and others used the rotational relaxation time, as originally defined by Peter Debye in his studies on dielectric phenomena. Only later (in the 1950s), during the development of nuclear magnetic resonance, was the term rotational correlation time used by Bloch. It thus seems reasonable for fluorescence practitioners to use ρ, but certainly adoption of either term should not lead to confusion. In terms of anisotropy and rotational correlation times, the Perrin equation becomes

$$\frac{r_o}{r} = \left(1 + \frac{\tau}{\tau_c}\right) \tag{5.13}$$

In the case of fluorescence probes associated noncovalently with proteins (e.g., porphryins, FAD, NADH, or ANS to give but a few systems), the probe is held to the protein matrix by several points of attachment (e.g., due to hydrogen bonds, ionic interactions, van der Waal interactions) and hence, its "local" mobility, that is, its ability to rotate independent of the overall "global" motion of the protein, is very restricted. In contrast, however, in the case of a probe attached covalently to a protein, via a linkage through an amine or sulfhydryl groups, for example (which will be discussed in Chapter 10), or in the case of tryptophan or tyrosine sidechains, considerable "local" motion of the fluorophore can occur. In addition, the protein may consist of flexible domains, which can rotate independent of the overall "global" protein rotation. This type of mobility hierarchy is illustrated in **Figure 5.18** for the case of a probe covalently attached to a dimeric protein. Where the various rotational modalities are (a) the overall rotation of the dimer, (b) the movement of one domain

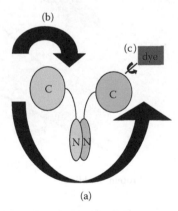

FIGURE 5.18 Illustration of different possible rotational modalities for the case of a fluorophore linked covalently to a protein. (Reprinted with permission from D.M. Jameson and J.A. Ross, 2010. *Chem. Rev.* 110: 2685–2708. Copyright 2010 American Chemical Society.)

independent of the overall protein rotation, and (c) the movement of the fluorophore about its point of attachment to the protein. All of these motions will contribute to the observed polarization of the fluorophore.

For many years, the presence and extent of such "local probe mobility" was judged by the use of Perrin–Weber plots, in which the viscosity of the solvent was varied either by the addition of sucrose or glycerol, followed by determination of the polarization after each addition. The data were then normally plotted as the reciprocal of the polarization minus one-third versus the temperature divided by the viscosity, as illustrated in **Figure 5.19** (top), for the case of protoporphyrin IX associated with horseradish peroxidase. In this case, the probe is tightly bound noncovalently in the heme binding site and there is no local probe mobility. **Figure 5.19** (bottom) simulates data for the case of a probe showing local as well as global motion (solid line) as well as a probe showing only one mode of mobility (dashed line).

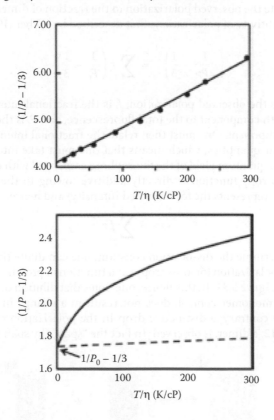

FIGURE 5.19 (Top) Perrin-Weber plot for protoporphyrin IX associated with Horseradish Peroxidase. (Adapted from Jullian et al., 1989. *Biochim. Biophys. Acta* 997: 206.) (Bottom) Simulation of Perrin-Weber plot for a free fluorophore and a fluorophore linked to a protein. (Reprinted with permission from D.M. Jameson and J.A. Ross, 2010. *Chem. Rev.* 110: 2685–2708. Copyright 2010 American Chemical Society.)

Some Applications of Polarization/Anisotropy
Protein Oligomerization

Polarization methods are ideally suited to study the aggregation state of a protein. Consider, for example, the case of a protein's dimer–monomer equilibrium. Following either intrinsic protein fluorescence (if possible) or by labeling the protein with a suitable probe, one would expect the polarization of the system to decrease upon dissociation of the dimer into monomers, since the smaller monomers will rotate more rapidly than the dimers, during the excited state lifetime (**Figure 5.20**).

Hence, for a given probe lifetime, the polarization (or anisotropy) of the monomer will be less than that of the dimer. In the concentration range near the dimer/monomer equilibrium constant, one expects to observe a polarization intermediate between that associated with either dimer or monomer. One can relate the observed polarization to the fraction of dimer or monomer using the additivity of polarizations first described by Weber (1952), namely:

$$\left(\frac{1}{<P>} - \frac{1}{3} \right)^{-1} = \sum f_i \left(\frac{1}{P_i} - \frac{1}{3} \right)^{-1} \tag{5.14}$$

where $<P>$ is the observed polarization, f_i is the fractional intensity contributed by the ith component to the total fluorescence, and P_i is the polarization of the ith component. One must then relate the fractional intensity contributions to molar quantities, which means that one must take into account any change in the quantum yield of the fluorophore associated with either species.

The anisotropy function is directly additive (owing to the fact that the denominator represents the total emitted intensity) and hence,

$$<r> = \sum f_i r_i \tag{5.15}$$

So, to determine the dissociation constant, one can dilute the protein and observe the polarization (or anisotropy) as a function of protein concentration as shown in **Figure 5.21**. In this figure, one notes that dilution of FITC-labeled lysozyme, a monomer control, does not result in a change in the polarization. On the contrary, a distinctive drop in the polarization of fluorescein-labeled L7/L12, a dimer, is observed. In fact the "span" (by span we mean 90%

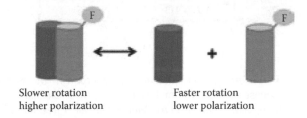

Slower rotation
higher polarization

Faster rotation
lower polarization

FIGURE 5.20 Depiction of a monomer-dimer equilibrium and the faster rotation (hence lower polarization) of the monomer compared to the dimer.

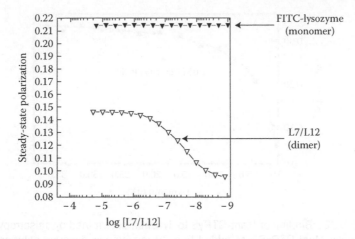

FIGURE 5.21 Plots of polarization versus concentration for a fluorescein labeled monomeric protein (lysozyme) compared to a fluorescein labeled dimeric protein (L7/L12). The dimer to monomer dissociation of the L7/L12 is evident from the decrease in polarization upon dilution. (Data from Hamman et al., 1996. *Biochemistry* 35: 16680. Reprinted with permission. Copyright 1996 American Chemical Society.)

dimer to 10% dimer) of the L7/L12 dissociation curve is 2.86 log units, which is expected for a dimer to monomer dissociation, just as the span of a simple ligand–protein interaction is 1.91 log units.

Protein/Ligand Interactions

The polarization/anisotropy approach is also very useful to study protein–ligand interactions in general. The first such study was carried out by David J. R. Laurence in 1952, working with Gregorio Weber's original polarization instrument in Cambridge, England. Laurence studied the binding of a series of fluorescent molecules, including xanthene, acridine, and naphthalene dyes, to bovine serum albumin. Although many probes do not significantly alter their quantum yield upon interaction with proteins, one should not take this fact for granted and one would be well advised to check. If the quantum yield does, in fact, change, one can readily correct the fitting equation to take this yield change into account. In terms of anisotropy, the expression relating observed anisotropy (r) to fraction of bound ligand (x), bound anisotropy (r_b), free anisotropy (r_f), and the quantum yield enhancement factor (g) is

$$x = \frac{r - r_f}{r_b - r_f + (g - 1)(r_b - r)} \tag{5.16}$$

In terms of polarization, the analogous equation is

$$x = \frac{(3 - P_b)(P - P_f)}{(3 - P)(P_b - P_f) + (g - 1)(3 - P_f)(P_b - P)} \tag{5.17}$$

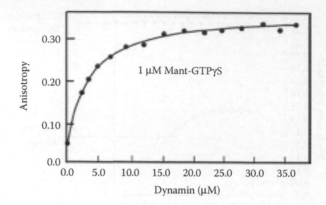

FIGURE 5.22 Binding of Mant-GTPγS to dynamin determined by anisotropy. One micromolar Mant-GTPγS was added to a 38 micromolar dynamin solution. The anisotropy was measured as the sample underwent successive dilutions with one micromolar Mant-GTPγS.

Note that the "g" used for the enhancement factor should not be confused with the "G-factor" discussed later in this chapter. A typical plot of polarization versus ligand/protein ratio is shown in **Figure 5.22**.

In this experiment, 1 μM mant-GTPγS (a fluorescent, very slowly hydrolyzable GTP analog) was present and the concentration of the GTP-binding protein, dynamin, was varied by starting with highly concentrated dynamin followed by dilution with buffer containing 1 μM mant-GTPγS. The binding curve was a fit to the anisotropy equation—in this case, the yield of the fluorophore increased by about twofold upon binding, and a K_d of 8.3 μM was found.

Proteolytic Processing

Proteolytic processing, mediated by proteolytic enzymes, or proteases, is critical to many vital biological processes, including posttranslational protein processing, blood clotting, digestion, hormone processing, apoptosis, and many others, as well as many deleterious processes, such as those mediated by anthrax and botulinum neurotoxins.

Hence, an evaluation of protease activity is often a requirement for an understanding of a particular pathway or for development of therapeutic agents. Protease assays have been around for many decades, in recent years, the development of rapid and sensitive protease assays suitable for high-throughput screening has attracted considerable attention. Fluorescence polarization lends itself very well to such assays since the essential aspect of a protease is to cleave a peptide bond, which almost always results in smaller molecular weight species. Hence, if the target protein can be labeled with a fluorescence probe, one would expect the polarization to decrease after proteolysis since the fluorophore will be able to rotate more rapidly after the protein mass to which it is tethered is reduced in size, as shown in **Figure 5.23**.

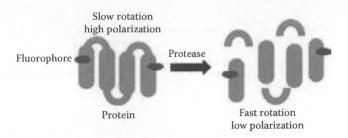

Slow rotation
high polarization

Fluorophore — Protease

Protein — Fast rotation
low polarization

FIGURE 5.23 Depiction of the change in polarization as a fluorophore-labeled protein is treated with a protease.

In addition to these types of polarization assays which follow proteolytic processes by taking advantage of the increased rotational mobility of the fluorophores after proteolysis, assays may also take advantage of the fact that depolarization can occur after Förster resonance energy transfer, FRET. This phenomenon will be discussed in some detail in Chapter 8.

Fluorescence Polarization Immunoassay

One of the most widely used techniques in clinical assays for drugs or metabolites is the fluorescence polarization immunoassay (FPIA). The application of fluorescence polarization to study antigen–antibody interactions was first developed by Walter Dandliker and his colleagues in the early 1960s. In this method, the target molecule, which is often but not always a drug or metabolite, is chemically linked to a fluorophore (the probe fluorescein is often used). In the presence of an antibody specific for the target molecule, the fluorophore-linked target will be bound to the antibody and its rotational mobility will be slow (relative to the fluorescence lifetime) and hence the emission will have a high polarization. When the appropriate sample is introduced, derived, for example, from blood or other fluids, the target molecule, if present, will bind to the antibody displacing the fluorophore-linked target. Free in solution, this fluorophore-target molecule will have an increased rotational mobility (lower Debye rotational relaxation time) and hence a lower polarization. The steps in this assay are illustrated in **Figure 5.24**.

Hence, it is relatively straightforward to set up a standard curve relating the observed polarization to the concentration of the target molecule in the sample. The diagnostic field, in general, uses the unit known as "millipolarization" or "mP" which is the normal polarization times 1000; for example, a polarization of 0.200 corresponds to 200 mP units. The mP unit was first introduced by Abbott Laboratories in a paper describing the first commercial implementation of the FPIA method with the TDx instrument (Jolley et al, 1981; see Additional Reading). Fluorescence plate readers, which can measure both intensity and polarization, are now routinely used for high-throughput screening. At the time of this writing, fluorescence polarization plate readers

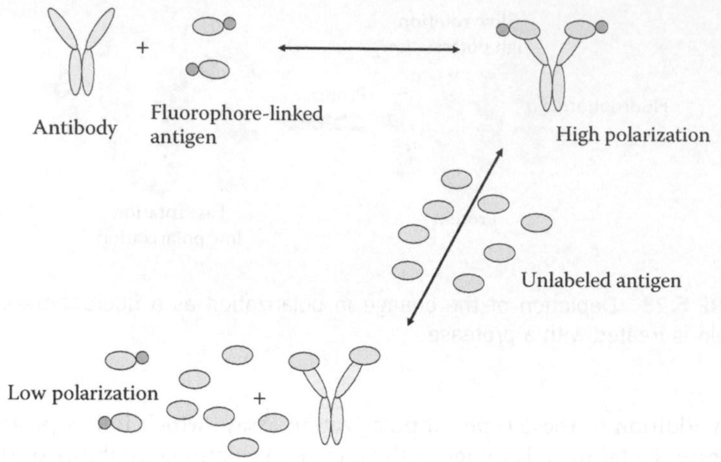

FIGURE 5.24 Depiction of the principles underlying a fluorescence polarization immunoassay (FPIA).

for 1536 wells are available—but one can expect that number to increase in the near future.

Membrane Fluidity

In 1971, Gregorio Weber and his colleagues introduced the use of polarization to study the physical state of lipids in model membrane systems. In that work they used the fluorescent dyes perylene (Figure 10.25), 9-vinylanthracene and 2-methylanthracene—probes which were chosen because of their favorable fluorescent properties including absorption and emission maxima as well as lifetimes and quantum yields. A few years later, Meir Shinitzky (who had been a postdoctoral fellow with Weber and who was a coauthor of the 1971 paper) and Yechezkel Barenholz introduced the probe diphenylhexatriene or DPH (Figure 10.25), which arguably became the most popular fluorescence polarization membrane probe of all time. As the structure indicates, DPH is nonpolar and will readily partition into lipid phases. The extent of rotational mobility of the DPH molecule, during the excited state lifetime, will depend upon the effective viscosity of the surrounding lipid phase. The term "fluidity" essentially represents the inverse of the effective viscosity. If the lipid phase is relatively rigid, for example, in bilayers composed of DMPC or DPPC at temperatures below their phase transitions of around 24°C and 40 V, respectively, then the polarization can be high—close to the limiting polarization for DPH. Raising the temperature, however, leads to a dramatic change in the physical state of the membrane—from the so-called ordered gel phase to the disordered liquid crystal phase, and the polarization decreases dramatically due to the increased mobility of the DPH. A typical DPH membrane experiment, showing the anisotropy of DPH in DPPC vesicles as a function of temperature, is shown in **Figure 5.25**.

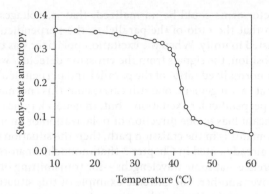

FIGURE 5.25 Anisotropy of DPH in DPPC vesicles as a function of temperature. (The author would like to thank Ivo Konopásek for this figure from his web site: http://web.natur.cuni.cz/~konop/gallery.php.)

Practical Aspects

G-Factor

To understand the so-called "G-factor," it is useful to think about polarization measurements carried out when I was a beginning graduate student in Gregorio Weber's laboratory. The instrument we used had been constructed by Weber as a T-format instrument, that is, with two detector arms symmetrically placed at 90° relative to the direction of the exciting light, as illustrated in **Figure 5.26**.

Imagine that the sample is excited by light polarized parallel to the vertical laboratory axis. The intensities measured along the two detector arms, which were viewed through either parallel or perpendicular polarizers, then should give the proper polarization according to Equation 5.1. The problem with this simple scenario, however, is that the two detector arms are not equal. One can easily imagine that one PMT may be more or less sensitive than the other, even if they are both operated at the same voltage. Also the details of the two optical arms are likely to differ slightly. To compensate for this lack of equivalency, we would first rotate the excitation polarizer to the perpendicular orientation. In this case, the polarization of the excitation is orthogonal to the transmission directions of the polarizers in the emission arms. Now

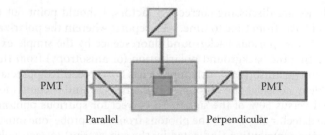

FIGURE 5.26 Schematic of a T-format polarization instrument.

the two detector arms could be normalized, that is, voltages to the PMTs adjusted such that the ratio of the parallel to the perpendicular intensities could be adjusted to unity. When the excitation polarizer was then rotated to the vertical position, the signals from the emission detectors would now give the correct or normalized ratios of the parallel and perpendicular intensities. In an "L format" arrangement, one still carries out this "normalization" procedure using perpendicular excitation—but, in general, typical PMTs do not exhibit significant bias to one direction of polarization or another. However, if a monochromator is in the emission path, then the situation is quite different. We have already noted in Chapter 3 that monochromators can exhibit a marked preference—at some wavelengths—for transmitting one direction of polarization over another. The extreme example of this situation, of course, is the Wood's anomaly. Hence, when the polarization is measured using a monochromator to isolate the emission—as opposed to an optical filter— then the correction factor can be significant. Although the normalization factor, in instruments wherein the emission was observed through an optical filter, was used by Weber in the 1940s and 1950s, and, in fact, by Francis Perrin and others in the 1920s and 1930s, it is nowadays usually referred to as the "G-factor," following a paper in 1962 by T. Azumi and S. P. McGlynn who carried out this procedure using a monochromator to isolate the emission—G-factor thus stands for "grating factor." Equations 5.18 and 5.19 illustrate how polarization values, corrected for the G-factor, are calculated. In Equation 5.18, the subscripts "V" and "H" refer to vertical and horizontal orientations of the excitation polarizer, respectively.

$$\frac{\left(\dfrac{I_\parallel}{I_\perp}\right)V}{\left(\dfrac{I_\parallel}{I_\perp}\right)H} = \left(\frac{I_\parallel}{I_\perp}\right)\text{Norm} \tag{5.18}$$

$$P = \frac{\left(\dfrac{I_\parallel}{I_\perp}\right)\text{Norm} - 1}{\left(\dfrac{I_\parallel}{I_\perp}\right)\text{Norm} + 1} \tag{5.19}$$

While we are discussing correction factors, I should point out that over the years I have, from time to time, seen reports wherein the polarization was "corrected" for spurious background fluorescence by the simple expedience of subtracting the background polarization (or anisotropy) from the sample polarization (or anisotropy). This approach is not correct! To appreciate merely one problem, consider the difference if the background accounts for 10% of the signal versus 50% of the signal. To correct for spurious photons, which reach the detector along with the photons from your probe, one must subtract the spurious contribution (indicated by the superscript) to each polarization component, as indicated in Equation 5.20.

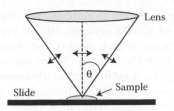

FIGURE 5.27 Illustration of the numerical aperture effect on polarization.

$$P_{corrected} = \frac{(I_\parallel - I_\parallel^s) - (I_\perp - I_\perp^s)}{(I_\parallel - I_\parallel^s) + (I_\perp - I_\perp^s)} \qquad (5.20)$$

Numerical Aperture Effects

Ideally, for the most accurate polarization measurements, one should use an excitation source with a very small angular dispersion as well as a small observation angle. In practice, one compromises by using lenses in the excitation and emission pathways which allow for reasonable illumination and light collection efficiencies. These lenses have associated with them a certain "cone angle" which is implied by the f-number of the lens, which is equal to f/D where f is the focal length and D is the diameter of the light cone (which usually corresponds to the lens diameter) (**Figure 5.27**). As pointed out by Weber (in a seminal 1956 paper which considered several sources of error in polarization instrumentation; see Additional Reading), the larger the numerical aperture of the lenses focusing the excitation and collecting the emission (i.e., the shorter the focal length and hence the larger the cone angle of collected light), the lower will be the measured polarization compared to the true polarization. This effect is, of course, exacerbated in fluorescence microscopy, wherein very large numerical apertures are often required to collect sufficient light. In normal, nonmicroscopic, research quality spectrofluorimeters, the deviation of the measured polarization from its true value is only a few percent—due to optics that typically subtend about 15°—but it is interesting to note that Weber's original instrument had less error since his optics used lenses which subtended only ~2.5°. The trade-off, of course, is larger light efficiency versus more accurate polarization values. In fluorescence microscopes, use of numerical apertures below 0.5 (typical for a 20× objective) leads to small errors, but for numerical apertures of 1.3 (typical for a 100× objective), the observed polarization can be ~20% below the true value.

Effect of Scattering

Rayleigh Scatter and Rayleigh Ghosts

Sometimes, samples exhibit turbidity due to large macromolecular assemblies, such as protein aggregates, or, in the case of membrane systems, vesicles. In such cases, one must consider the effect of scattering on polarization/

anisotropy measurements. Before considering the effects of scattering on the measurement *per se*, we should consider trivial, but common, instrument artifacts. The main concern in such cases is that some of the excitation light, Rayleigh scatter, will reach the detector. When one is dealing with a system that is highly scattering, one must be particularly careful with the choice of the emission filter or, if a monochromator is used on the emission side, one must be careful with the wavelength and the slitwidth. In cases wherein emission filters are used, a trick I learned from Gregorio Weber is useful. Before taking the measurement, one places the emission filter in the excitation path to verify that the signal then observed in the emission channel is not significantly above the dark level. If a significant signal is detected, then one must choose a different, hopefully more effective, emission filter. In some cases, though, one finds that no emission filter seems effective. In such cases, one may be dealing with parasitic light, as mentioned in Chapter 3 in the discussion on monochromators. Recall that parasitic light refers to light reaching the detector, which does not correspond to the target wavelengths, light often termed as Rayleigh ghosts. Normally, in optically clear solutions, this light is of little consequence since it is present at a low level and passes straight through the solution. If there is significant scattering, however, then some of this parasitic light can be directed toward the emission side and may pass through any longpass filter and reach the detector. In a polarization measurement, the scattering will raise the apparent polarization—the same as if direct Rayleigh scatter reached the detector. If there is parasitic light present, one can insert an interference filter—which passes the excitation wavelength—in the excitation path to "clean up" the exciting light (i.e., to exorcize the Rayleigh ghosts). It is common, in fact, for researchers who routinely work with samples giving rise to appreciable scattering—such as membrane systems—to utilize fluorimeters with a double monochromator in the excitation path. This arrangement has the advantage that the light incident on the sample is spectrally very pure, but the disadvantage that the intensity of the excitation will be significantly decreased compared to single-monochromator excitation.

Scattering of the Fluorescence

If parasitic light as well as Rayleigh scatter is eliminated, what effect will sample turbidity have on the polarization of the fluorescence? In fact, turbidity, which is due to multiple scattering, will always decrease the fluorescence polarization. This decrease occurs because both excitation and emission will be rotated as a consequence of the multiple scattering events. The magnitude of this effect can be significant. For example, John Teale showed that addition of 0.7% glycogen to a solution of dansyl labeled serum albumin, which increased the optical density of the solution at 366 nm from 0.02 to 0.3, caused a drop in the polarization from 0.306 to 0.250. One way to minimize this effect is to use smaller pathlength cuvettes, for example, 3 mm square instead of 1 cm square. I wish to emphasize that blank subtraction will not correct for this depolarization effect since the decrease is inherent in the measurement. Joseph Eisinger and Jorge Flores (1985, *Biophys. J.* 48:77), while studying the polarization of membrane probes in erythrocyte membranes, described how

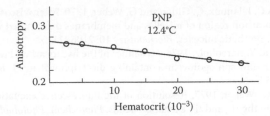

FIGURE 5.28 Anisotropy of 1-phenyl-3-(2-naphthyl)-2-pyrazoline (PNP) in red blood cells as a function of hematocrit. (Adapted from J. Eisinger and J. Flores, 1985. *Biophys. J.* 48: 77.)

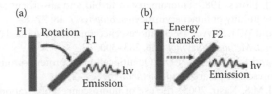

FIGURE 5.29 Depiction of depolarization due to (a) rotational diffusion and (b) energy transfer.

anisotropy measurements at decreasing hematocrits allowed one to extrapolate to an anisotropy value corrected for the effect of scattering (**Figure 5.28**).

Depolarization via FRET

An important mechanism of depolarization is transfer of the excited state energy, that is, FRET. As illustrated in **Figure 5.29**, rotational diffusion of an excited state dipole or transfer of excited state energy from one dipole to another of a different orientation, will both result in a change in the observed polarization of the emission. FRET theory and this phenomenon of depolarization via FRET will be discussed in Chapter 8.

Additional Reading

G. Weber, 1952. Polarization of the fluorescence of macromolecules I. Theory and experimental method. *Biochem. J.* 51: 145–155.

D.J.R. Laurence, 1952. A study of the adsorption of dyes on bovine serum albumin by the method of polarization of fluorescence. *Biochem. J.* 51: 168–180.

G. Weber, 1956. Photoelectric method for the measurement of the polarization of the fluorescence of solutions. *J. Opt. Soc. Am.* 46: 962–970.

G. Weber, 1966. Polarization of the fluorescence of solutions. In *Fluorescence and Phosphorescence Analysis*, pp. 217–240. Ed. D. Hercules. Interscience Publishers, New York.

M. Shinitzky, A.C. Dianoux, C. Gitler, and G. Weber, 1970. Microviscosity and order in the hydrocarbon region of micelles and membranes determined with fluorescent probes. I. Synthetic micelles. *Biochemistry* 10: 2106–2113.

M. Shinitzky and Y. Barenholz, 1976. Dynamics of the hydrocarbon layer in liposomes of lecithin and sphingomyelin containing dicetylphosphate. *J. Biol. Chem.* 249: 2652–2657.

B. Valeur and G. Weber, 1977. Resolution of the fluorescence excitation spectrum of indole into the ^1L_a and ^1L_b excitation bands. *Photochem. Photobiol.* 25: 441–444.

D.M. Jameson, G. Weber, R.D. Spencer, and G. Mitchell, 1978. Fluorescence polarization: Measurements with a photon-counting photometer. *Rev. Sci. Instru.* 49: 510–514.

M.E. Jolley, S.D. Stroupe, C.H. Wang, H.N. Panas, C.L. Keegan, R.L. Schmidt, and K.S. Schwenzer, 1981. Fluorescence polarization immunoassay. I. Monitoring amino-glycoside antibiotics in serum and plasma. *Clin. Chem.* 27: 1190–1197.

J. Eisinger and J. Flores, 1985. Fluorometry of turbid and absorbant samples and the membrane fluidity of intact erythrocytes. *Biophys. J.* 48: 77–84.

D.M. Jameson and W.H. Sawyer, 1995. Fluorescence anisotropy applied to biomolecular interactions. *Methods Enzymol.* 246: 283–300.

D.M. Jameson and S.E. Seifried, 1999. Quantification of protein–protein interactions using fluorescence polarization. *Methods* 19: 222–233.

M.E. Jolley and M.S. Nasir, 2003. The use of fluorescence polarization assays for the detection of infectious diseases. *Comb. Chem. High Throughput Screening* 6: 235–244.

D.M. Jameson and G. Mocz, 2005. Fluorescence polarization/anisotropy approaches to study protein–ligand interactions: Effects of errors and uncertainties. *Methods Mol. Biol.* 305: 301–322.

D.M. Jameson and J.A. Ross, 2010. Fluorescence polarization/anisotropy in clinical diagnostics and imaging. *Chem. Rev.* 110: 2685–2708.

M. Nazari, M. Kurdi, and H. Heerklotz, 2012. Classifying surfactants with respect to their effect on lipid membrane order. *Biophys J.* 102: 498–506.

6

Time-Resolved Fluorescence

Excited State Lifetimes

Knowledge of a fluorophore's excited state lifetime is crucial for quantitative interpretations of numerous fluorescence measurements such as quenching, polarization, and FRET. We should first ask, though, exactly what is meant by the "lifetime" of a fluorophore? Although we often speak of the properties of fluorophores as if they are studied in isolation, such is not usually the case. Absorption and emission processes are almost always studied on *populations* of molecules and the properties of the supposed typical members of the population are deduced from the macroscopic properties of the population. I note, however, that modern single-molecule techniques have allowed lifetime measurements on isolated, single fluorophores—which are achieved by exciting the fluorophores many times (usually on the order of thousands of times) and building up a histogram of the fluorescence decays analogous to the type of data obtained on bulk samples.

George Stokes, who as mentioned in Chapter 1 coined the term "fluorescence," was aware of the phenomenon of phosphorescence and understood that considerable time could elapse between the absorption of light and the phosphorescence phenomenon (a fact mentioned by Galileo)! The possibility of a similar, though much shorter, time lag in fluorescence occurred to Stokes and, though he expressed doubt that such a lag existed, he wrote: "I have not attempted to determine whether any appreciable duration could be made out by means of a revolving mirror." As we shall learn, this lag time is so short that poor Stokes would have had to rotate his mirror really, really fast!

One of the first serious attempts to measure fluorescence lifetimes was by R.W. Wood, who in 1922 described an ingenious experiment in which he shot streams of pressurized dyes, such as anthracene, through a nozzle designed so that he could illuminate the tip with a beam of light. The idea was that the dye would be excited at the nozzle's tip and then travel along in the stream before emitting fluorescence. Hence, the lifetime of the excited state could be estimated from the length of the fluorescent part of the stream. Even though Wood could resolve distances as short as 0.1 mm, he did not observe any fluorescence beyond the excitation point. Since he knew the velocity of the stream, ~200 m/s, he could calculate that the fluorescence lifetime must be less than 1/2,300,000 s, that is, less than ~435 ns. Although Wood failed to measure the time between absorption and emission, his work was important in that he was able to put an upper boundary to the fluorescence lifetime.

The first accurate lifetime measurements were made a few years later by Enrique Gaviola, one of the truly great experimental physicists of the early twentieth century, who modulated the intensity of the exciting light and carried out the first successful *frequency domain* measurement (this approach will be described in detail later). He obtained lifetimes of 2 ns for rhodamine and 4.5 ns for fluorescein, both of which are reasonable values. In the 1920s, Francis Perrin was also able to use polarization data to calculate fairly accurate lifetime values for several dyes with short lifetimes, such as eosin and erythrosin.

In general, the behavior of an excited population of fluorophores is described by a familiar rate equation:

$$\frac{dn^*}{dt} = -n^*\Gamma + f(t) \tag{6.1}$$

where n is the number of excited elements at time t, Γ is the rate constant of emission, and $f(t)$ is an arbitrary function of the time, describing the time course of the excitation. The dimensions of Γ are s^{-1} (transitions per molecule per unit time). If excitation occurs at $t = 0$, the last equation, takes the form:

$$\frac{dn^*}{dt} = -n^*\Gamma \tag{6.2}$$

and describes the decrease in excited molecules at all further times. The integration gives

$$n^*(t) = n^*(0)e^{-\Gamma t} \tag{6.3}$$

The lifetime, τ, is equal to Γ^{-1}. If a population of fluorophores is excited, the lifetime is the time it takes for the number of excited molecules to decay to $1/e$ or 36.8% of the original population according to

$$\frac{n^*(t)}{n^*(0)} = e^{-t/\tau} \tag{6.4}$$

as shown in **Figure 6.1**, in pictorial form.

Time Domain

Excited state lifetimes have traditionally been measured using either the *impulse* response or the *harmonic* response method. In principle, both methods have the same information content. These methods are also referred to

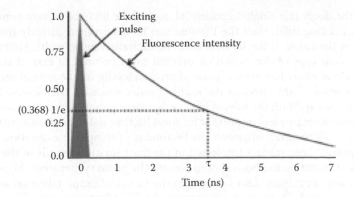

FIGURE 6.1 Depiction of intensity decay plot showing a short excitation pulse and a fluorescence signal.

as the "time domain" method and the "frequency domain" method. In the *impulse* (or time domain) method, the sample is illuminated with a short pulse of light and the intensity of the emission versus time is recorded. Originally these short light pulses were generated using *flashlamps*, which had widths on the order of several nanoseconds. Modern laser sources can now routinely generate pulses with widths on the order of picoseconds or shorter.

As shown in **Figure 6.1**, the fluorescence lifetime, τ, is the time at which the intensity has decayed to $1/e$ of the original value. The decay of the intensity with time is given by the relation:

$$I_t = \alpha e^{-t/\tau} \tag{6.5}$$

where I_t is the intensity at time t, α is a normalization term (the preexponential factor), and τ is the lifetime. It is more common to plot the fluorescence decay data using a logarithmic scale as shown in **Figure 6.2**.

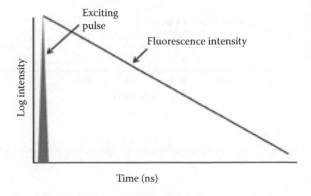

FIGURE 6.2 Logarithmic display of Figure 6.1.

If the decay is a single exponential, and if the lifetime is long compared to the exciting light, then the lifetime can be determined directly from the slope of the curve. If the lifetime and the excitation pulse width are comparable, some type of *deconvolution* method must be used to extract the lifetime. Great effort has been expended on developing mathematical methods to "deconvolve" the effect of the exciting pulse shape on the observed fluorescence decay. With the advent of very fast laser pulses, these deconvolution procedures became less important for most lifetime determinations, although they are still required whenever the lifetime is of comparable duration to the light pulse. Those readers interested in the mathematical details of the available deconvolution methods should consult the primary literature. **Figure 6.3** shows intensity decay data for anthracene in cyclohexane taken on an IBH time correlated single photon counting (TCSPC) instrument (more on this technique later). One sees that these data correspond to a single exponential decay (note the straight line which extends over several decades of intensity) and that the number of time channels between excitation and the $1/e$ point of the intensity decay is ~76. Given that there are 56 ps/channel, that means that the lifetime is 76 times 0.056 ns or 4.1 ns. The curve below the intensity decay plot is the chi-square plot, which represents an evaluation of the goodness-of-fit between the data and a decay model (more on this later). One should also note that the signal appears to have more noise as the time gets longer. This increase in noise is to be expected since there are fewer and fewer photons

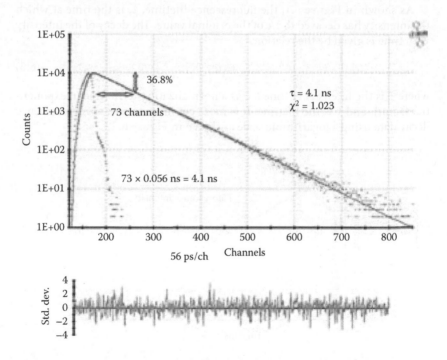

FIGURE 6.3 Intensity decay (lifetime) of anthracene in ethanol (from IBH).

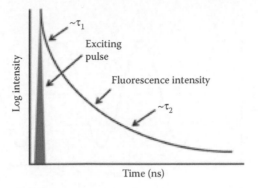

FIGURE 6.4 Depiction of logarithmic display of two lifetime components.

present at longer times, and since the noise in such a signal will be proportional to the square root of the number of photons.

Unfortunately, the majority of lifetimes determined on biological systems, such as proteins or membranes, cannot be fit to single exponential decays. The origin of these complex decay kinetics may be simply due to mixtures of single exponential decaying components or it may be due to subtle (and sometimes not so subtle) interactions of the fluorophore with its environment. In some obvious cases, such as multi-tryptophan proteins, fluorophores are located in different environments and lifetime heterogeneity is expected. If the decay is multiexponential, the relation between the intensity and time after excitation is given by

$$I(t) = \sum_i \alpha_i e^{-t/\tau_i} \tag{6.6}$$

One may then observe data such as those sketched in **Figure 6.4**.

Here, we can discern at least two lifetime components indicated as τ_1 and τ_2. This presentation is oversimplified but illustrates the point.

Frequency Domain

In the harmonic method (also known as the phase and modulation or frequency domain method), a continuous light source is utilized, such as a laser or xenon arc, and the intensity of this light source is modulated sinusoidally at high frequency as depicted in **Figure 6.5**. Typically, an *electro-optic* device, such as a *Pockels cell* is used to modulate a continuous light source, such as a CW laser or a xenon arc lamp. Alternatively, light sources such as LEDs or laser diodes can be directly modulated.

In such a case, the excitation frequency is described by

$$E(t) = E0 \, [1 + M_E \sin \omega t] \tag{6.7}$$

$E(t)$ and $E0$ are the intensities at time t and 0, M_E is the modulation factor, which is related to the ratio of the AC (alternating part) and DC (average part)

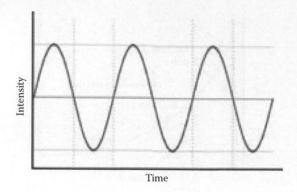

FIGURE 6.5 Depiction of sinusoidal waveform of exciting light in frequency domain fluorometry.

components of the signal, and ω, the angular modulation frequency, equals $2\pi f$, where f is the linear modulation frequency. Due to the persistence of the excited state, fluorophores subjected to such an excitation will give rise to a modulated emission, which is shifted in phase relative to the exciting light as depicted in **Figure 6.6**.

Figure 6.6 illustrates the phase delay (ϕ) between the excitation, $E(t)$, and the emission, $F(t)$. Also shown are the AC and DC levels associated with the excitation and emission waveforms. One can demonstrate that:

$$F(t) = F0 \left[1 + M_F \sin(\omega t + \Phi)\right] \qquad (6.8)$$

This relationship indicates that measurement of the phase delay, ϕ, can form the basis of one measurement of the lifetime, τ. In particular, as demonstrated in 1933 by F. Dushinsky:

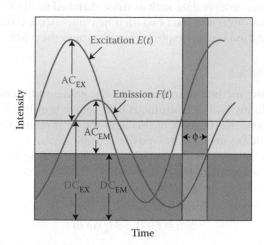

FIGURE 6.6 Illustration of phase shift and relative modulation measurements.

$$\tan \Phi = \omega\tau \tag{6.9}$$

The *modulations* of the excitation (ME) and the emission (MF) are given by

$$M = \left(\frac{AC}{DC}\right)_E \quad \text{and} \quad M = \left(\frac{AC}{DC}\right)_F \tag{6.10}$$

The *relative modulation*, M, of the emission is then

$$M = \frac{(AC/DC)_F}{(AC/DC)_E} \tag{6.11}$$

τ can also be determined from M according to the relation:

$$M = \frac{1}{\sqrt{1 + (\omega\tau)^2}} \tag{6.12}$$

Using the *phase shift (i.e., phase delay)* and *relative modulation (i.e., modulation ratio)*, one can thus determine a *phase lifetime* (τ_P) and a *modulation lifetime* (τ_M). Multifrequency phase and modulation data are usually presented as shown in **Figure 6.7**. The plot shows the frequency–response curves, both phase and modulation, for anthracene in ethanol acquired on a ChronosFD (ISS, Inc., Champaign, IL) using a 370 nm LED. The emission was collected through a WG389 longpass filter. The data are best fitted by a single exponential decay time of 4.25 ns.

If the fluorescence decay is a single exponential, τ_P and τ_M will be equal at all modulation frequencies. However, if the fluorescence decay is multiexponential, then $\tau_P < \tau_M$ and, moreover, the values of both τ_P and τ_M will depend upon the modulation frequency, specifically, both the phase and modulation lifetimes will decrease as the excitation light modulation frequency increases, that is,

$$\tau_P(\omega_1) > \tau_P(\omega_2) \quad \text{if } \omega_1 < \omega_2$$

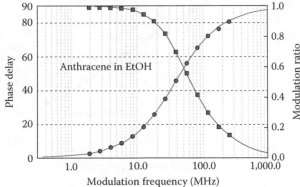

FIGURE 6.7 Phase and modulation results for anthracene in ethanol. (The author would like to thank ISS, Inc. for this figure.)

To get a feeling for typical phase and modulation data, consider the following data set.

Frequency (MHz)	τ_P (ns)	τ_M (ns)
5	6.76	10.24
10	6.02	9.70
30	3.17	6.87
70	1.93	4.27

These differences between τ_P and τ_M, and their frequency dependence, form the basis of the methods used to analyze for lifetime heterogeneity, that is, the component lifetimes and amplitudes. In the case just shown, the actual system being measured was a mixture of two fluorophores with lifetimes of 12.08 ns and 1.38 ns, with relative contributions to the total intensity of 53% and 47%, respectively. Here, we must be careful to distinguish the term *fractional contribution to the total intensity* (usually designated as f) from α, the preexponential term referred to earlier. The relation between these two terms is given by

$$f_i = \frac{\alpha_i \tau_i}{\sum_j \alpha_j \tau_j} \tag{6.13}$$

where j represents the sum of all components. In the case just given, the ratio of the preexponential factors corresponding to the 12.08 and 1.38 ns components is approximately 1/8. In other words, there are about eight times as many molecules in solution with the 1.38 ns lifetime as there are molecules with the 12.08 ns lifetime. **Figure 6.8** shows frequency domain data for

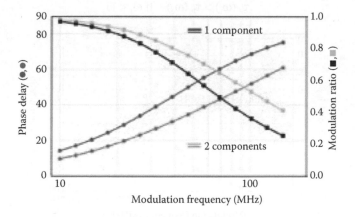

FIGURE 6.8 Simulation of phase and modulation data for a one component system ($\tau = 4.05$ ns) and a two component system ($\tau_1 = 4.05$ ns, $f_1 = 0.5$; $\tau_2 = 1.0$ ns, $f_2 = 0.5$). (The author would like to thank Carissa Vetromile for this figure.)

both one ($\tau = 4$ ns) and two ($\tau_1 = 4$ ns, $f_1 = 0.5$; $\tau_2 = 1$ ns, $f_2 = 0.5$) component systems.

Data Analysis

In addition to decay analysis using discrete exponential decay models, one may fit the data to *distribution* models. In this case, it is assumed that the excited state decay characteristics of the emitting species actually results in a large number of lifetime components. **Figure 6.9** shows a typical lifetime distribution plot for the case of single tryptophan containing protein—human serum albumin. The distribution shown is Lorentzian, but depending on the system, different types of distributions, for example, Gaussian or asymmetric distributions, may be utilized. This approach to lifetime analysis is described in Alcala et al., 1987 *Biophys. J.* 51: 597. **Figure 6.10** depicts both a distribution analysis and a two-component discrete analysis for mant-GDP bound to N-Ras P21 protein. In this case, both fits give approximately the same chi-square value (see below). So one may ask—which is the correct analysis? Well, that is the rub, as they say. One cannot always decide, based on lifetime analysis alone, what type of approach is best to describe the underlying photophysics of a system. The fact is that one often needs more information, for example, biochemical, chemical, or spectroscopic, to decide how best to approach a particular case.

In the case of either discrete or distribution models, the appropriateness of the model is judged by the reduced chi-square (χ^2) value which is given by

$$\chi^2 = \sum \frac{[(P_c - P_m)/\sigma_P]^2 + [(M_c - M_m)/\sigma_M]^2}{2n - f - 1} \tag{6.14}$$

where P and M refer to phase and modulation data, respectively; the subscripts c and m refer to calculated and measured values; σ_P and σ_M refer to the standard deviations of each phase and modulation measurement, respectively; n

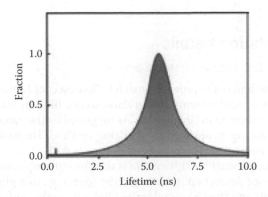

FIGURE 6.9 Example of Lorentzian distribution lifetime analysis for HSA. (The author would like to thank Leonel Malacrida for this figure.)

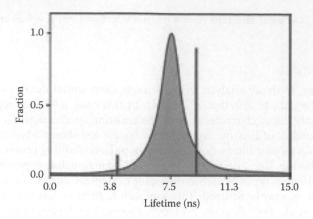

FIGURE 6.10 Comparison of discrete (red) and distribution (blue) lifetime analyses for mant-GDP bound to N-Ras protein P21. Both approaches give similar chi-square values. (Modified from D.M. Jameson and T.L. Hazlett, 1991. In *Biophysical and Biochemical Aspects of Fluorescence Spectroscopy*. Ed. G. Dewey, Plenum Press, New York, pp. 105–133.)

is the number of modulation frequencies and f is the number of free parameters. As an aside, I note that the precision of modern commercial phase and modulation instruments is generally around 0.2° of phase angle and 0.004 modulation values (or better) and these values are typically used in the chi-squared calculations. "Good" fits between the data and the model (i.e., single exponential, two exponential, etc.) typically are characterized by reduced chi-squared values around 1.

Another popular lifetime analysis method—based on information theory—is the *maximum entropy method* (MEM). In this method, no *a priori* intensity decay model is assumed. For an excellent treatment of this approach see the review by Jean-Claude Brochon (1994) listed in Additional Reading.

Instrumentation Details

Time Correlated Single Photon Counting

The TCSPC method is a popular approach to fluorescence lifetime determinations, for both *in vitro* systems and in fluorescence lifetime imaging microscopy (FLIM) (discussed in Chapter 10). As suggested by the name, this method is a photon counting, as opposed to an analog, method. The basic components of TCSPC instrumentation are depicted in **Figure 6.11**.

The electronic "heart" of this system is the time-to-amplitude converter or TAC. This device detects an incoming pulse (due, e.g., to a photon from the excitation light) and then a second pulse (due, e.g., to the emission) and converts the time duration between the arrival of these pulses into a voltage. This output voltage, in turn, can be used to address a particular memory location

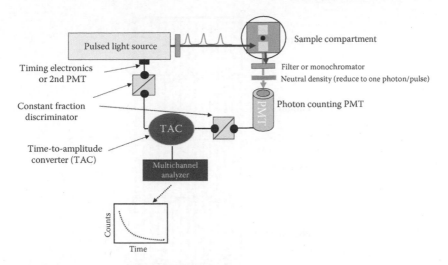

FIGURE 6.11 Sketch of time correlated single photon counting (TCSPC) instrumentation.

or channel in a "multichannel analyzer," which then registers a count in that channel. When the TAC detects another pair of pulses, another channel (or if the time delay is identical to the first pulse, the same channel) registers a count. As more pulse pairs are detected, a histogram of counts versus time builds up in the multichannel analyzer. **Figure 6.12** illustrates this process.

Clearly, the precision of the lifetime determination will ultimately depend upon the total number of counts being analyzed. For a particular precision, for example, ±0.1 ns, one may need several thousand total counts. More counts are required, however, if the task is to resolve multiple lifetimes, since the precision of a resolved component will depend upon the number of photons that particular component contributes to the total intensity decay. Certain caveats apply to TCSPC data collection. For one, the sample rate must be sufficiently low such that pulse pileup does not occur, leading to reduction of apparent counts in some channels and biasing the lifetime.

Streak Cameras

Streak cameras can provide intensity, wavelength, and lifetime information. Essentially, a streak camera transforms the temporal profile of a light pulse into a spatial profile on a detector. Although streak cameras can be based on a rotating mirror system (such as those used for very high-speed photography), in the fluorescence field, streak cameras are usually based on purely electronic systems. In these types of streak cameras, incident photons hit a photocathode, which produces photoelectrons (just as in PMTs). These electrons are accelerated and pass through an electric field that deflects them according to their time of arrival. This operation produces a streak (hence the name) which carries temporal and intensity information. These devices are inherently very, very

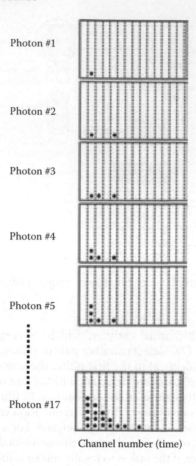

Photon #1

Photon #2

Photon #3

Photon #4

Photon #5

Photon #17

Channel number (time)

FIGURE 6.12 Depiction of histograms of accumulating photon counts for emission decay in the time-domain.

fast and resolutions down to subpicoseconds can be achieved. In addition to the temporal resolution, streak cameras can be combined with spectroscopes to also achieve wavelength resolution. A typical streak camera image is illustrated in **Figure 6.13**.

Frequency Domain Instrumentation

A classic instrumentation approach to phase and modulation measurements is depicted in **Figure 6.14**. In this example, the light from a continuous light source, such as a CW laser or a xenon arc lamp, is modulated sinusoidally using a Pockels cell, an electro-optic device. Nowadays, instrumentation is also available which uses intrinsically modulated sources, such as LEDs or laser diodes. The careful reader will note that in **Figure 6.14**, the two synthesizers (S1 and S2) are not running at the same frequency. Specifically, S2 is

FIGURE 6.13 A streak camera image. The image, showing the emission decay of horseradish peroxidase A2 excited at 280 nm, illustrates the temporal and spatial resolution provided by this method. (Modified from Neves-Peterson et al., 2007. *Biophys J.* 92: 2016.)

running at the S1 frequency plus an additional 400 Hz. This additional 400 Hz is called the cross-correlation frequency and is used to modulate the dynode chain of the PMTs. The idea, used with Weber's original cross-correlation phase and modulation fluorometer (Spencer and Weber, 1969; see Additional Reading), was to transfer all the phase and modulation information to the

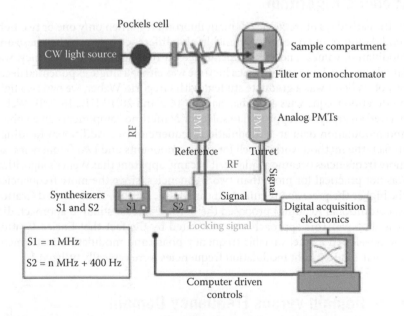

FIGURE 6.14 Sketch of frequency domain instrument utilizing a CW light source and a Pockels cell as the light modulator.

low-frequency domain, since it was much easier to work with an audio signal than with a megahertz signal. Of course with this heterodyning procedure, there is also a signal at the base megahertz frequency plus 400 Hz, but this is easily filtered out electronically. *Anecdote Alert! During my early days as a graduate student in Gregorio Weber's laboratory, I used to work with the original cross-correlation instrument, which could operate at 14.2 and 28.4 MHz. The cross-correlation frequency at that time was 36 (or 72) Hz, generated by a rotating capacitor. The light modulation was accomplished using a Debye–Sears ultrasonics tank, which had the RF crystal set up at one end of the tank and a stainless-steel reflector at the other end. When properly aligned, a standing wave of density fluctuations was established in a solution of water/ethanol. To change frequencies, one had to drain the tank and replace the radio-frequency crystal. This procedure took some hours, since one also had to carefully align the reflector to optimize the standing waves and hence the light modulation. Aligning this reflector took skill and patience and I remember good-natured competitions with Bernard Valeur, a postdoc at that time in Weber's lab, to see who could get the highest light modulation value. Bernard usually won!*

Another approach is to use the harmonic content of pulsed sources. For example, Alcala et al. (1985) (see Additional Reading) utilized the harmonic content of a mode-locked laser to acquire multifrequency phase and modulation data. In fact, the time and frequency domain are related by Fourier transformations, but a detailed discussion of the theory is outside the scope of this book.

Weber's Algorithm

In the early days of frequency domain fluorometry, when only one or two light modulation frequencies were available, the difference between the phase and modulation values, and their dependence on the modulation frequency, was largely used to demonstrate that a lifetime was either a single exponential decay or not. When I was a graduate student with Gregorio Weber, we had two light modulation frequencies available, namely 14.2 and 28.4 MHz. In 1981, Weber derived an analytical approach to solve for N lifetime components given phase and modulation data at N modulation frequencies (see Additional Reading). In fact, the method worked well for two components and two frequencies. As more frequencies became available, it became apparent that Weber's algorithm was not practical for more than two frequencies, since the more frequencies, the higher the precision required. At that point, in 1983, Jameson and Gratton (see Additional Reading) proposed the nonlinear least-squared approach discussed above. This approach was motivated by the fact that Enrico Gratton had developed a true, variable frequency phase and modulation instrument, and that 10 to 20 light modulation frequencies were typically utilized.

Time Domain versus Frequency Domain

I am sometimes asked which technique, time domain or frequency domain, is better. I can attest to the fact that in the early days, the arguments between the time and frequency domain aficionados were often harsh (which is an understatement)! I am reminded of Jonathan Swift's novel *Gulliver's Travels*. At one point, Gulliver describes a dispute between the residents of two island nations, Lilliput and Blefuscu, who wage war over the question of whether a soft-boiled egg should be broken at the little end or the big end (the proponents of each approach were called either Little-endians or Big-endians). Jonathan Swift's point was that the argument had grown all out of proportion. In theory, the information content of both time domain and frequency domain methods are virtually identical. I say "virtually," since time-domain measurements invariably involve truncation of the data range at both short and long times, whereas all lifetime components contribute to all frequencies measured. There are some practical considerations which affect each method, but in my experience, some people are just more comfortable with one approach or the other. In 2007, a study was published in which nine laboratories, some using time-domain and some using frequency domain approaches, all carried out lifetime determinations on a series of fluorophores (see Boens et al. (2007) in Additional Reading). In fact, regardless of the method utilized, all labs found very similar lifetime values—lifetimes ranged from 89 ps to 31.4 ns, depending on the fluorophore and the solvent. Comprehensive tables compiling the results for numerous fluorophore/solvent combinations were presented. Some of the results are reproduced in **Table 6.1**.

Magic Angles

Jablonski was the first to point out that rotational diffusion of fluorophores could result in apparent changes in fluorescence lifetimes. Specifically, he showed that the intensity decay of the parallel polarized component would be shorter than the true molecular intensity decay, while the intensity decay of the perpendicular component would be longer than the true molecular decay. To understand this point, consider the usual right-angle observation of an excited fluorophore. If the sample is excited with light polarized parallel to the observation axis, then the observed fluorescence intensity decreases due to two effects, namely, (1) the normal decline in intensity due to the emitting population of excited states and (2) the rotation of the excited dipole away from the parallel observation direction. In essence, the constraints of right angle observation, which allows one to observe the dipoles emitting parallel and perpendicular to the observation direction—but not the dipoles oriented along the observational axis—introduces an observational bias into the system. To correct for this bias, one can use polarizers oriented at the "magic angle." With right angle observations, four combinations of excitation/emission polarizers fulfill magic angle conditions. These magic angle conditions are: (1) 45° excitation–35° emission, (2) 35° excitation–no polarizer emission, (3) 0° (parallel) excitation–55° emission, (4) 55° excitation–0° (parallel) emission. The derivation of these angles is beyond the scope of this book; readers seeking more information can refer to Weber and Spencer (1970) in Additional Reading.

Table 6.1 Comparison of Lifetimes Obtained Using Frequency Domain and Time Domain Approaches			
Fluorophore	Solvent	FD Lifetime (ns)	TD Lifetime (ns)
Anthracene	MeOH	5.00	5.20
	Cyclohexane	5.32	5.32
9-Cyanoanthracene	MeOH	15.29	16.27
	Cyclohexane	12.39	13.47
9,10-Diphenylanthracene	MeOH	8.71	8.77
	Cyclohexane	7.17	7.76
N-methylcarbazole	Cyclohexane	14.06	14.15
Coumarin 153	MeOH	4.18	4.33
Erythrosine B	Water	0.090	0.089
	MeOH	0.45	0.48
NATA	Water	3.14	3.01
POPOP	Cyclohexane	1.12	1.12
PPO	MeOH	1.63	1.66
	Cyclohexane	1.35	1.38
Rhodamine B	Water	1.73	1.75
	MeOH	2.48	2.44
Rubrene	MeOH	9.79	9.97
N-(3-sulfopropyl) acridinium	Water	30.90	31.37
p-Terphenyl	MeOH	1.10	1.20
	Cyclohexane	0.96	1.00

Anisotropy Decay/Dynamic Polarization

In Chapter 5, I considered steady-state polarization/anisotropy methodologies and I mentioned that when a probe is covalently linked to a protein, it may undergo local mobility, independent of the overall or global rotation of the protein. To resolve these mobilities, Weber used the approach of varying the viscosity of the solution using sucrose or glycerol. An example of this type of Weber–Perrin plot (as it is known) was shown in Figure 5.19. With the development of time domain measurements, another approach to separation of molecular mobilities became possible. Time-resolved polarization/anisotropy methodologies provide information on changes in fluorophore

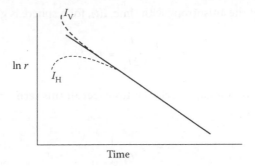

FIGURE 6.15 Depiction of time-decay anisotropy data. The dotted lines correspond to the intensity decay of the vertical (I_V) and horizontal (I_H) polarized emission components while the solid line corresponds to the log of the anisotropy ($\ln r$).

orientation as a function of time. The time domain approach is usually termed the *anisotropy decay* method, while the frequency domain approach is known as *dynamic polarization*. In principle, both methods yield the same information.

In the time domain anisotropy method, the sample is illuminated by a short pulse of vertically polarized light and the decay over time of both the vertical and horizontal components of the emission are recorded. The anisotropy function is then plotted versus time as illustrated in **Figure 6.15**. Note that the horizontal component actually increases during short times, since initially the fluorophores, which started out more or less aligned in the parallel direction, have not rotated significantly, meaning that there are few dipoles oriented in the horizontal direction. As time passes, though, the number of horizontally oriented molecules increase.

Let us consider how to treat different types of systems. Consider first the simplest case, that of a spherical body. In this case, we assume that the fluorophore has no local mobility—such is the case for noncovalent interactions (**Figure 6.16**).

FIGURE 6.16 Depiction of fluorophores bound non-covalently to a spherical macromolecule and the rotation of the system.

The decay of the anisotropy with time, $I(t)$, for a sphere is given by

$$r = \frac{I_{\|} - I_{\perp}}{I_{\|} + 2I_{\perp}} = r_0 e^{-t/\tau_c} \tag{6.15}$$

where τ_c is the rotational correlation time (recall this term from Chapter 5), and where:

$$\tau_c = \frac{1}{6D_{\text{rotation}}} \tag{6.16}$$

In the case of nonspherical particles, such as ellipsoids, the time decay of anisotropy function is more complicated (**Figure 6.17**).

Mathematically simple symmetrical ellipsoids give us a sense of how changes in an object's shape affect its rotational diffusion rates. In the case of symmetrical ellipsoids of revolution, the relevant expression for the time decay of anisotropy is

$$r(t) = r_1 e^{-t/\tau_{c1}} + r_2 e^{-t/\tau_{c2}} + r_3 e^{-t/\tau_{c3}} \tag{6.17}$$

where

$$\tau_{c1} = \frac{1}{6D_2}$$

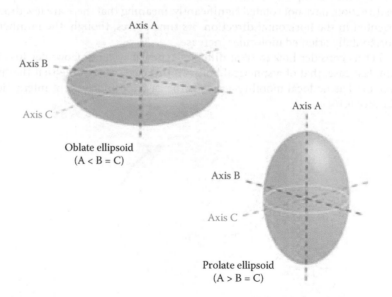

Oblate ellipsoid
(A < B = C)

Prolate ellipsoid
(A > B = C)

FIGURE 6.17 Depiction of the principle rotational axes of prolate and oblate ellipsoids of revolution.

$$\tau_{c2} = \frac{1}{(5D_2 + D_1)}$$

$$\tau_{c3} = \frac{1}{(2D_2 + 4D_1)} \tag{6.18}$$

D_1 and D_2 are the rotational diffusion coefficients about the axes of symmetry and about either equatorial axis, respectively. Resolution of the rotational rates is limited in practice to two rotational correlation times, which differ by at least a factor of 2. The anisotropy amplitudes (r_1, r_2, and r_3) relate to orientation of the probe with respect to the axes of symmetry for the ellipsoid (**Figure 6.18**) *(we are assuming colinear excitation and emission dipoles)*.

$$r_1 = 0.1(3\cos^2\phi - 1)^2$$

$$r_2 = 0.3\sin^2(2\phi) \tag{6.19}$$

$$r_3 = 0.3\sin^4(\phi)$$

What about the case of a mixture of rotating species? Mixed systems, as shown in **Figure 6.19** for the case of a mixture of large and small particles, show simple, multiexponential behavior; the anisotropy decay reflects the sum of the exponential components present, as indicated in Equation 6.19.

$$r(t) = r_1 e^{-t/\tau_{c1}} + r_2 e^{-t/\tau_{c2}} \tag{6.20}$$

Consider the case of a "local" rotation of a probe attached to a spherical particle (**Figure 6.20**).

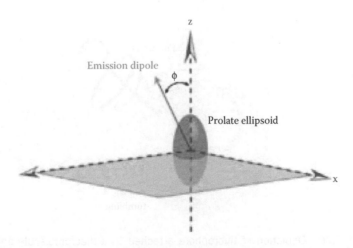

FIGURE 6.18 Depiction of the axes of a prolate ellipsoid of revolution with an emission dipole oriented at an angle to the long axis.

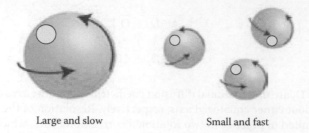

Large and slow   Small and fast

FIGURE 6.19 Depiction of a mixture of two types of rotating spherical particles: large particles with slow rotation and small particles with fast rotation.

In this case, the expression for the decay of anisotropy is

$$r(t) = r_1 e^{-t/\tau_{c1}} + r_2 e^{-(t/\tau_{c1} + t/\tau_{c2})} \tag{6.21}$$

where τ_{c1} represents the "global" probe motion and τ_{c2} represents the "local" rotation of the macromolecule. Such designations are clearly an oversimplification, but they are a useful convention, as long as one keeps in mind that the actual rotational modalities are probably more complicated.

In some systems, a fluorophore experiences hindered rotation. For example, a fluorophore in a membrane may only be able to rotate to a limited extent, depending on the physical state of the bilayer (**Figure 6.21**). This scenario is sometimes referred to as "wobble-in-a-cone." In these cases, the final anisotropy level reached is termed "r-infinity" or r_∞. The value of r_∞ depends on the angular displacement, Φ, reached, or

$$\frac{r_0}{r_\infty} = \frac{3\cos^2 \Phi - 1^2}{2} \tag{6.22}$$

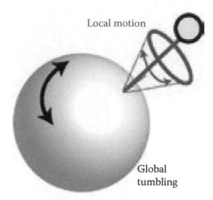

Local motion

Global
tumbling

FIGURE 6.20 Depiction of fluorophore attached to a macromolecule and able to undergo rotational motion (local motion) independent of the rotation of the macromolecule (global motion).

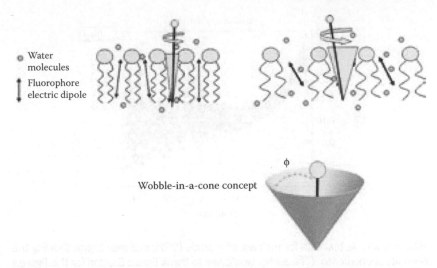

Water molecules

Fluorophore electric dipole

Wobble-in-a-cone concept

ϕ

FIGURE 6.21 Illustration of limited rotational mobility of a membrane association fluorophores and the dependence of the extent of rotation on the physical state of the lipid bilayer. The concept of "wobble in a cone" is also illustrated.

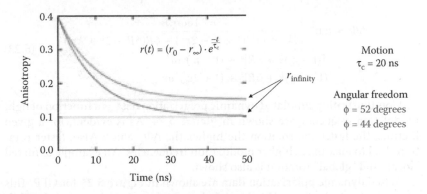

$$r(t) = (r_0 - r_\infty) \cdot e^{\frac{-t}{\tau_c}}$$

r_{infinity}

Anisotropy

Time (ns)

Motion
$\tau_c = 20$ ns

Angular freedom
$\phi = 52$ degrees
$\phi = 44$ degrees

FIGURE 6.22 Illustration of the anisotropy decay data to be expected from the case illustrated in Figure 6.21, where the limited mobility of the fluorophores (during its fluorescence lifetime) leads to a limiting non-zero anisotropy termed "r infinity."

These principles are illustrated in **Figure 6.22**. Actual data for DPH in a membrane is shown in **Figure 6.23**. In this figure, one notes that the scatter in the data increases at longer times, when there are fewer emission events.

Dynamic Polarization

In dynamic polarization measurements, the sample is illuminated with vertically polarized, modulated light. The phase delay between the parallel and

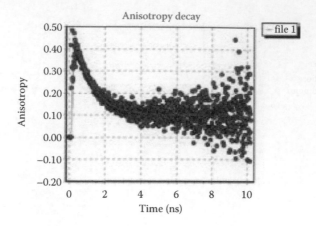

FIGURE 6.23 Actual data for the case of a probe (DPH) in a membrane showing the r-infinity phenomenon. (The author would like to thank Enrico Gratton for this figure.)

perpendicular components of the emission is measured, as well as the modulation ratio of the AC contributions of the perpendicular and parallel components. For a spherical particle, Weber derived the equations:

$$\Delta\Phi = \tan^{-1}\left[\frac{18\omega r_0 R}{(k^2 + \omega^2)(1 + r_0 - 2r_0^2) + 6R(6R + 2k + kr_0)}\right]$$

$$Y^2 = \frac{[(1 - r_0)k + 6R]^2 + (1 - r_0)^2\omega^2}{[(1 - 2r_0)k + 6R]^2 + (1 + 2r_0)^2\omega^2}$$

(6.23)

Plots illustrating simulated dynamic polarization data as a function of light modulation frequency are shown in **Figure 6.24**. As is evident, for a given lifetime, the faster the rotation the higher the $\Delta\Phi$ values. Also, faster rotations lead to maxima at higher modulation frequencies. An example of mixed "local" and "global" rotation is also shown.

Actual dynamic polarization data are shown in **Figure 6.25** for GFP. This system is clearly dominated by the global protein rotation; that is, local mobility of the fluorophores is virtually absent (as one would expect for the chromophore in GFP).

Figure 6.26 shows dynamic polarization results obtained on a single tryptophan system, namely, elongation factor Tu bound to GDP and also complexed with elongation factor Ts (which has no tryptophan residues). These data were obtained using 300 nm excitation, which ensure the highest polarization value. The TuGDP data indicate that the tryptophan residue experiences very little local mobility. Upon binding with Ts, however, it is immediately apparent that this tryptophan residue experiences greatly enhanced local mobility. Interestingly, the steady-state polarization of TuGDP was higher than that of TuTs (0.270 compared to 0.216). This result at first may seem counter-intuitive, since TuGDP is significantly smaller than TuTs (43 kDa compared to 74 kDa). Without the time-resolved data, one might speculate that the lifetime of the

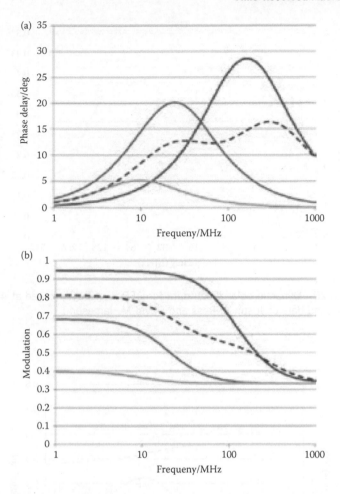

FIGURE 6.24 Differential phase data (a) and modulation ratio (b) for an isotropic rotator with 3 ns (solid red), 30 ns (solid blue) and 300 ns (solid green) rotational relaxation times. The dashed blue line in the phase and modulation curves correspond to the case of two rotational relaxation times, namely 30 ns and 1.5 ns, with associated anisotropies of 0.2 for each component. In each case a fluorescence lifetime of 20 ns was used and colinear excitation and emission dipoles (*i.e.*, limiting anisotropies of 0.4) were assumed. (Adapted from J.A. Ross and D.M. Jameson, 2008. *J. Photochem. Photobiol. Sci.* 7: 1301 with permission from the European Society for Photobiology, the European Photochemistry Association, and The Royal Society of Chemistry.)

tryptophan residue is much longer in TuTs than in TuGDP, to account for the increased depolarization. In fact, the lifetimes are virtually the same in both cases, and the lower polarization in TuTs is due to the greater local tryptophan mobility. This example nicely illustrates the value of time-decay anisotropy data.

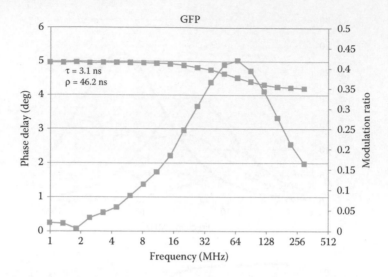

FIGURE 6.25 Dynamic polarization data for GFP in at 22C, excited at 471 nm. (Unpublished data of Justin A. Ross and David M. Jameson.)

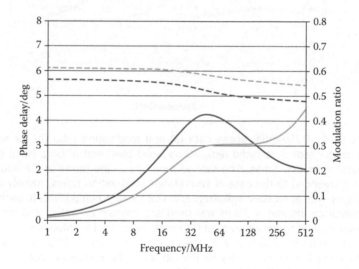

FIGURE 6.26 Differential phase data (solid) and modulation ratio (dashed) for EF-Tu-GDP (red) ($\tau_1 = 4.8$ ns, $\tau_2 = 0.31$ ns, $f_1 = 0.79$, $\rho_1 = 63$ ns, $r_1 = 0.23$, $\rho_2 = 1.5$ ns, $r_2 = 0.05$) and EF-Tu-EF-Ts (green) ($\tau_1 = 4.6$ ns, $\tau_2 = 0.23$ ns, $f_1 = 0.82$, $\rho_1 = 84$ ns, $r_1 = 0.19$, $\rho_2 = 0.3$ ns, $r_2 = 0.12$). (Adapted from J.A. Ross and D.M. Jameson, 2008. *J. Photochem. Photobiol. Sci.* 7: 1301 with permission from the European Society for Photobiology, the European Photochemistry Association, and The Royal Society of Chemistry.)

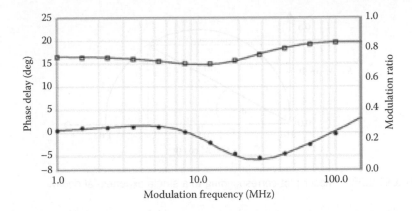

FIGURE 6.27 Dynamic polarization data for a mixture of ethidium bromide (EB) and yeast transfer RNA (phase delay—solid circles; modulation ratio—open squares). The EB concentration is about 4-fold higher than the tRNA concentration. (The author would like to thank Carissa Vetromile for this figure.)

Cases in which there are contributions from both bound and free fluorophores, that is, mixtures of fast and slow rotations, can give rise to odd looking data, as illustrated in **Figure 6.27**. The system shown is yeast transfer RNA in the presence of an excess of ethidium bromide. Some of the ethidium bromide is bound to the tRNA and has a long lifetime (~27 ns) and a slow rotational relaxation time (~84 ns), while some has a short lifetime (~1.8 ns) and a fast rotational relaxation time (~1.5 ns). As shown, in such cases the $\Delta\Phi$ values can dip below zero. Since I first saw this phenomenon in the data of my postdoc, Theodore "Chip" Hazlett, I named it the "Chip Dip." One also sees a related phenomenon in time domain decay of anisotropy data, in which the anisotropy decreases at first and then rises up before decreasing again. In this case, it is easy to understand that the initial anisotropy is due to the short lifetime/fast rotation component and that after that component has decayed, the long lifetime/slow rotation component takes over. In the time domain, this phenomenon is called the "dip and rise."

Phasors

A relatively recent addition to the armamentarium of the fluorescence practitioner is the phasor. As I shall describe at the end of this section, phasor plots can actually be calculated using either frequency domain or time domain data, but I shall first illustrate them using frequency domain results. A phasor diagram is essentially a simple geometrical representation of the time-resolved data. The basic concept is as follows. Given phase (Φ) and modulation (M) values at a particular light modulation frequency, two values, termed S and G, can be calculated by the expressions: $G = M \cos \Phi$ and $S = M \sin \Phi$ (these parameters were first described by Gregorio Weber). Using these G and

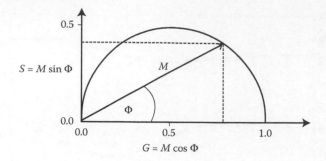

FIGURE 6.28 Phasor plot corresponding to a single exponential decay.

S values, one can construct a plot as shown in **Figure 6.28**. These types of plots have been called phasor, polar, and AB plots. In this book I shall use the designation "phasor plots."

The perimeter of this curve is known as the "universal circle," and all single exponential lifetimes must fall somewhere on this circle. The exact location of the point will depend upon the light modulation frequency. If we now consider the case of an emission arising from two fluorophores, each a single exponential decay, then the individual components in this system would each lie on the universal circle but their mixture must result in G and S values that lie on the line connecting the two values on the universal circle, as shown in **Figure 6.29**.

Figure 6.30 shows the results obtained for a mixture of three single exponential components—both binary and ternary mixtures are shown. In this figure, the data were obtained using 20 MHz modulation. As shown in **Figure 6.31**, the curves move when the modulation frequency is changed (the reader will recall from earlier in this chapter that higher frequencies weight the shorter lifetime components so that the phase and modulation values alter with modulation frequency).

In fact, these phasor diagrams provide a useful way to visualize complex decays. For example, if we consider a protein having numerous tryptophan residues, it can be very challenging to demonstrate the effects of ligand binding or protein unfolding. With the phasor plot, however, no matter how many

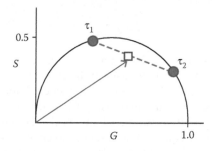

FIGURE 6.29 Phasor plot corresponding to a system containing two discrete lifetime components.

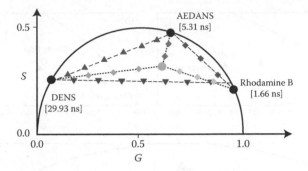

FIGURE 6.30 Phasor plots taken at 20 MHz light modulation frequency for three different fluorophores and for different mixtures of these fluorophores. Binary mixtures give rise to phasor points along the line connecting the pure fluorophores while ternary mixtures result in phasor points inside the triangle formed by the three 2-component mixtures.

tryptophans contribute to the emission, at a given light modulation frequency there will be only one phasor point. If ligand binding alters the excited state properties of one or more emitting residues, the phasor point will move, as illustrated in **Figure 6.32**, for the cases of GDP and GTP binding to dynamin (a) and thyroxine and furosemide binding to HSA (b). Hence, movement of the phasor point can be an indication of a conformational change in the protein. **Figure 6.32**c demonstrates that if two proteins do not interact, for example, lysozyme (closed square) and antithrombin (solid circle), the phasor point due to their mixture (open square) must lie along the line between the phasor points for each individual protein. If the proteins interact, however, then it is likely that one or more tryptophan residues of one or both proteins will have altered lifetime properties, with the result that the phasor point corresponding to their mixture will not lie on the line connecting the two individual protein phasor points. This scenario is shown for the case of antithrombin

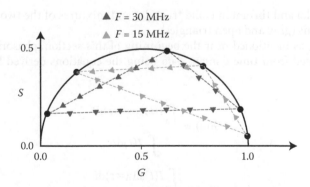

FIGURE 6.31 The same system shown in Figure 6.30 illustrating the effect of changing the light modulation frequency on the overall position of the phasor points.

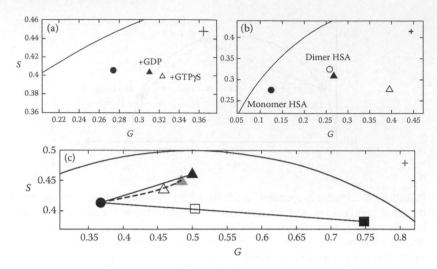

FIGURE 6.32 (A) Phasor plot for dynamin 2 alone (circle), GDP-bound dynamin (closed triangle), and GTPγS-bound dynamin (open triangle). (B) Monomeric HSA (closed circle) plotted with furosemide-bound HSA (open triangle) and D-thyroxine-bound HSA (closed triangle). Dimeric HSA (open circle), which can be found at levels up to 10% in lyophilized HSA, has a unique decay/phasor point compared with monomeric HSA. (C) Phasor plot showing protein–protein interaction. Pure antithrombin (solid circle), pure lysozyme (closed square), and a mixture of both proteins (open square) all lie on a straight line (connecting each pure protein phasor point), indicating the lack of protein interaction. Thrombin is represented by a solid triangle and mixtures of antithrombin/thrombin (gray triangle 0.5:1; open triangle 1:1) give rise to phasor points that are clearly off the line connecting the two pure proteins, indicating protein interaction. Crosshairs seen in the upper part of each figure represent the statistical error for each phasor point on the individual phasor plot scales. (Adapted from N.G. James et al., 2011. *Anal. Biochem.* 410: 62.)

(solid circle) and thrombin (solid triangle) and mixtures of the two, interacting proteins (gray and open triangles).

Finally, as mentioned near the beginning of this section, phasors can also be calculated from time domain data using the equations derived by Weber, namely,

$$G(\omega) = \frac{\int_0^\infty I(t)\cos(\infty t)\mathrm{d}t}{\int_0^\infty I(t)\mathrm{d}t}$$

$$S(\omega) = \frac{\int_0^\infty I(t)\sin(\infty t)\mathrm{d}t}{\int_0^\infty I(t)\mathrm{d}t}$$

(6.24)

In these cases, one may choose any convenient frequency, for example, the repetition rate frequency of the exciting light source (e.g., 80 MHz for many Ti–sapphire two photon laser systems used for FLIM).

Additional Reading

J.R. Alcala, E. Gratton, and D.M. Jameson, 1985. A multifrequency phase fluorometer using the harmonic content of a mode-locked laser. *Anal. Instrum.* 14: 225–250.

N. Boens, W. Qin, N. Basarić, J. Hofkens, M. Ameloot, J. Pouget, J.P. Lefèvre et al., 2007. Fluorescence lifetime standards for time and frequency domain fluorescence spectroscopy. *Anal Chem.* 79: 2137–2149.

J.-C. Brochon, 1994. Maximum entropy method of data analysis in time-resolved spectroscopy. *Methods Enzymol.* 240: 262–311.

E. Gratton and D.M. Jameson, 1985. New approach to phase and modulation resolved spectra. *Anal. Chem.* 57: 1694–1697.

E. Gratton, D.M. Jameson, N. Rosato, and G. Weber, 1984. A multifrequency cross-correlation phase fluorometer using synchrotron radiation. *Rev. Sci. Instrum.* 55: 486–494.

N.G. James, J.A. Ross, M. Stefl, and D.M. Jameson, 2011. Application of phasor plots to *in vitro* protein studies. *Anal. Biochem.* 410: 70–76.

D.M. Jameson, E. Gratton, and R.D. Hall, 1984. The measurement and analysis of heterogeneous emissions by multifrequency phase and modulation fluorometry. *Appl. Spectros. Rev.* 20: 55–106.

D.M. Jameson and T.L. Hazlett, 1991. Time-resolved fluorescence measurements in biology and biochemistry. In *Biophysical and Biochemical Aspects of Fluorescence Spectroscopy*, pp. 105–133. Ed. G. Dewey, Plenum Press, NY.

D.M. Jameson and G. Weber, 1981. Resolution of the pH-dependent heterogeneous fluorescence decay of tryptophan by phase and modulation measurements. *J. Phys. Chem.* 85: 953–958.

J.A. Ross and D.M. Jameson, 2008. Frequency domain fluorometry: Applications to intrinsic protein fluorescence. *J. Photochem. Photobiol. Sci.* 7: 1301–1312.

R.D. Spencer and G. Weber, 1969. Measurement of subnanosecond fluorescence lifetimes with a cross-correlation phase fluorometer. *Anal. NY Acad. Sci.* 158: 361–376.

R.D. Spencer and G. Weber, 1970. Influence of Brownian rotations and energy transfer upon the measurements of fluorescence lifetime. *J. Chem. Phys.* 52: 1654–1663.

M. Stefl, N.G. James, J.A. Ross, and D.M. Jameson, 2011. Application of phasor plots to *in vitro* time-resolved fluorescence measurements. *Anal. Biochem.* 410: 62–69.

G. Weber, 1981. Resolution of the fluorescence lifetimes in a heterogeneous system by phase and modulation measurements. *J. Phys. Chem.* 85: 949–953.

Quantum Yields and Quenching

IN THIS CHAPTER, I WILL discuss two "Q" topics, namely quantum yields and quenching. Both of these topics have a long pedigree. The fluorescence quantum yield (QY) was a concept introduced in 1924 by Sergey Ivanovich Vavilov. The pivotal paper on fluorescence quenching appeared in 1919, authored by Otto Stern and Max Volmer. Let us start with QY.

Perhaps the simplest definition of QY is that it represents the number of photons emitted divided by the number of photons absorbed—as indicated in Equation 7.1

$$QY = \text{photons emitted as fluorescence/photons absorbed} \qquad (7.1)$$

This equation refers to the usual case of steady-state illumination, that is, the sample is illuminated by a constant intensity light source, and the actual intensity being measured corresponds to fluorescence per unit time. So, if a fluorophore absorbs 100 photons and emits 100 photons, its QY is 1.0 or 100% whereas if it absorbs 100 photons and emits only 20 photons, then its QY is 0.20 or 20%. We cannot define the QY as the energy emitted divided by the energy absorbed since, given the Stoke's shift, the emission is virtually always of lower energy than the absorbed light. Hence, if we considered the energies involved, even a perfect emitter would have a QY less than 1, and moreover the QY would change as the excitation wavelength varied. A more insightful definition of QY would be that the QY equals the *rate of the emission process* divided by the sum of the *rates of all other deactivation processes*.

$$QY = k_f / \Sigma k_d \qquad (7.2)$$

where k_f is the rate of fluorescence, and Σk_d designates the sum of the rate constants for the various processes, in addition to fluorescence, that depopulate the excited state. These nonradiative deactivation processes may include photochemical and dissociative processes in which the products are well-characterized chemical species (electrons, protons, radicals, molecular isomers). Also, we may have less well-characterized changes that result in a return to the ground state with the simultaneous dissipation of the excited state energy as heat. These latter processes are collectively called "radiationless transitions," and two types have been clearly recognized: intersystem crossing and internal conversion. Intersystem crossing refers to the radiationless spin inversion of one electron in the excited singlet state, S_1, which results in the isoenergetic, or almost isoenergetic, conversion into a triplet state. When spin inversion occurs in S_1, the resulting triplet reached is T_1, which lies in an intermediate energy position between S_1 and S_0. In the case of internal conversion, the molecular spin state remains the same, and the energy is given off as vibrational modes and hence converted into heat. The QY is, in most cases, independent of the exciting wavelength, though there are some rare exceptions (but discussion of these cases are beyond the scope of this book) and one can speak without ambiguity of the QY of a pure fluorophore after specifying the solvent composition and temperature.

Determination of QYs

Despite the occasional optimistic report to the contrary, QYs are notoriously difficult to determine accurately—the key word being "accurately." In support of this point of view, I offer the fact that "reliable" QYs for quinine sulfate from "excellent" laboratories range from approximately 0.50–0.70. Fortunately, absolute QYs are rarely required for most fluorescence studies. In the great majority of cases, one only needs to follow changes in the fluorescence yield of a target system, as the result of some chemical or physical process. If you are determined to accurately measure the QY of a fluorophore, I will try to point you in the right direction.

QYs are usually determined using either "absolute" or "relative" methods. In an "absolute" method, the sample is typically enclosed in an integrating sphere (**Figure 7.1**). The fluorescing sample is placed in the sphere's interior, which is coated with a highly reflective material such that the emitted light will reflect many times, but will eventually reach the detector. Hence, essentially all of the emitted light can be measured and quantified relative to a highly scattering sample. One can find some absolute QY determinations in the literature, such as the classic work by Weber and Teale in 1956. The Hamamatsu Corporation recently introduced an instrument for absolute QY determinations, which uses the integrating sphere approach (the C9920-12 External Quantum Efficiency System). However, the majority of QY determinations are carried out using "relative" methods, which entail the use of a standard and comparison of the yield of the sample to that of the standard. The concept is easy enough, but as I remarked earlier, the devil is in the details.

The classic approach for determination of a QY using the "relative" method is to select a fluorophore, a standard, which has absorption and emission properties that roughly match those of your sample. For example, tryptophan or N-acetyltryptophanamide (NATA) are common standards for proteins, while quinine sulfate and fluorescein are often used when the excitation is in

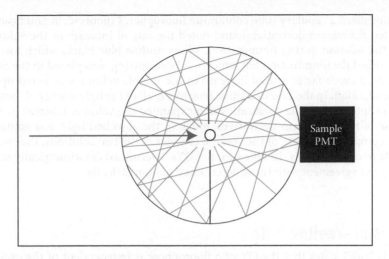

FIGURE 7.1 Sketch of an integrating sphere used for QY determinations. The exciting light (blue) impinges on the sample, in the center of the sphere, which emits fluorescence (green) in all directions. This emission reflects off the internal surface and eventually reaches the detector (sample PMT). (The author would like to thank Leonel Malacrida for this figure.)

the mid-UV or visible regions, respectively. Then, one makes up solutions of the standard and sample such that the optical densities are similar. Now, one simply measures the emission spectra of both standard and sample, applies correction factors to get true molecular spectra, and integrates the area under the spectra to get the total intensity of the emitted light. One then uses the following equation to calculate the QY of the sample:

$$QY_{sample} = QY_{standard} \, (I_{T\,sample}/I_{T\,standard})(1{-}10^{-ODstandard}/1{-}10^{-ODsample})$$
$$(n_{sample}/n_{standard})^2 \tag{7.3}$$

The term $(1{-}10^{-OD})$ represents the fraction of light absorbed at the excitation wavelength. I should note that, in the literature, one often sees these terms replaced with absorbances or optical densities, which is only a good approximation if the optical densities are very low. The I_T terms are, of course, the total intensities (more about these terms soon), and n refers to the refractive index. When publishing QYs, it is important to describe the methodology in detail and especially to indicate the QY assumed for the standard. That way when a skeptical reader (like me) happens along, he or she can substitute his or her own preferred value for the QY of the standard.

Interestingly, one does not have to use optical spectroscopy to determine a QY. In some cases one can utilize calorimetry. The basic concept is that the excited state can be depopulated by nonradiative pathways, which compete with the light emission process. When nonradiative pathways are active, though, this energy goes into the surroundings and thus raises the temperature of the solvent. Seybold et al. (see Additional Reading), for example, illuminated a capillary tube containing fluorophores (fluorescein and brominated fluorescein derivatives), and noted the rate of increase in the volume of the solution during illumination. When aniline blue black, which totally absorbed the illumination but which was nonemitting, was placed in the capillary, a much faster rate of increase in the sample volume was noted upon illumination. In the case of aniline blue black, all of the light energy absorbed went into heating the solution, which expanded the volume, whereas in the case of the fluorescein derivatives, much of the absorbed light was emitted. By comparing the rate of the volume changes of the two solutions, they were able to calculate QYs. In all cases, the QYs determined calorimetrically were in good agreement with those determined photometrically.

Kasha–Vavilov Rule

This "rule" states that the QY of a fluorophore is independent of the excitation wavelength. Generally, this rule holds since, as stated earlier, regardless of the excitation wavelength the emission is from the lowest vibrational level of the first excited electronic state. There are occasional exceptions to this rule, however. For example, the QYs of indole and tryptophan vary depending on whether they excited into the lowest energy electronic states (S_1) or the higher energy electronic levels (S_2). The yield of tryptophan, for example, begins to drop as the excitation wavelength decreases below 240 nm, reaching a value about 50% lower than that upon excitation above 240 nm. The reason for this decreased QY is presumably electron ejection from the higher electronic excited state. Similar observations have been made for benzene and naphthalene derivatives. The lifetime of tryptophan has, on the other hand, been shown to be independent of the excitation wavelength—down to 200 nm—which means that excited tryptophans that are not quenched by electron ejection undergo the normal decay process.

Quenching

A number of chemical and/or physical processes can lead to a reduction in fluorescence intensity. Such intensity reductions are termed "quenching." In this chapter, we shall mainly discuss cases involving reversible quenching due to the molecular interaction, that is, contact of the fluorophore, in either its ground or excited state, with another molecule. Other mechanisms that can lead to a loss of emission from the fluorophore include energy transfer, charge transfer reactions or photochemistry. We shall focus our attention on the two quenching processes most commonly utilized by the fluorescence practitioner—namely collisional (dynamic) quenching and static (complex formation) quenching. We may first ask, though, why does one do these types of experiments, that is, what information can we gain? In the majority of cases, quenching provides information on the accessibility of the fluorophore to the solvent. This accessibility may depend not only on structural features of the fluorophore's surroundings but its charge environment as well. Later in this chapter we shall see specific examples of quenching experiments, which will illustrate this point.

Collisional or Dynamic Quenching

Collisional quenching occurs when the excited fluorophore experiences contact with an atom or molecule that can facilitate nonradiative transitions to the ground state. Common quenchers include O_2, I^-, and acrylamide. **Figure 7.2** shows a solution of fluorescein, in the absence and presence of iodide ion. In the simplest case of collisional quenching, the following relation, called the *Stern–Volmer equation*, holds:

$$F_0/F = 1 + K_{SV}[Q] \qquad (7.4)$$

where F_0 and F are the fluorescence intensities observed in the absence and presence, respectively, of quencher, $[Q]$ is the quencher concentration and K_{SV} is the Stern–Volmer quenching constant. A plot of F_0/F versus $[Q]$ should thus yield a straight line with a slope equal to K_{SV}. Such a plot, known as a Stern–Volmer plot, is shown in **Figure 7.3** for the case of fluorescein quenched by iodide ion (I^-).

In this case, $K_{SV} \sim 8.3$ L mol^{-1}. K_{SV} is equal to $k_q\tau_0$ where k_q is the quenching rate constant, and τ_0 is the excited state lifetime in the absence of quencher. The quenching rate constant, k_q, is equal to the quenching efficiency, γ, times the diffusion-limited bimolecular rate constant. In many cases, γ is assumed to be unity and hence, k_q can be taken as the bimolecular quenching rate constant, which takes into account the sum of the diffusion rates and molecular radii of both fluorophore and quencher. Specifically,

$$k_q = 4\pi aDN' \qquad (7.5)$$

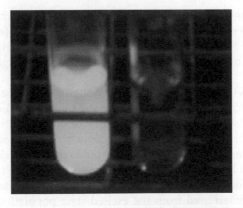

FIGURE 7.2 Photo of fluorescein solution, illuminated using a UV handlamp, in the absence (left) and presence (right) of potassium iodide. (From Croney et al., 2001. *Bio Chem.* and *Mol. Biol. Ed.* 29: 600.)

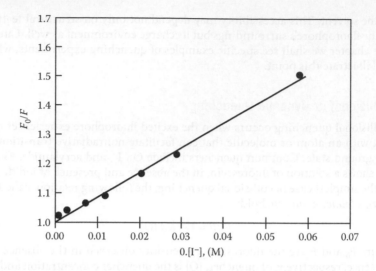

FIGURE 7.3 Stern–Volmer plot for quenching of fluorescein by iodide ion.

where N' is Avogadro's number divided by 1000 (i.e., the number of molecules per millimole), a is the sum of the molecular radii of fluorophore and quencher, and D is the sum of their diffusion coefficients. Although this assumption that γ is unity is reasonable for many cases, its validity depends upon the precise quenching mechanism. For example, some quenchers operate via an electron spin exchange mechanism while others utilize electron transfer processes or spin–orbital coupling mechanisms. A detailed examination of quenching mechanisms is beyond the scope of this book and I shall present the topic in a more phenomenological manner. In the case of purely collisional (dynamic) quenching:

$$F_0/F = \tau_0/\tau \qquad (7.6)$$

Hence, in this case: $\tau_0/\tau = 1 + k_q\tau_0[Q]$ (a relationship which only holds if the lifetime is monoexponential). In the fluorescein/iodide system, $\tau_0 = 4$ ns and $k_q \sim 2 \times 10^9$ M^{-1} s^{-1}. One may ask "Why does dynamic quenching reduce the lifetime?" The concept underlying this phenomenon is roughly illustrated in **Figure 7.4.**

The reader will recall, from Chapter 6, that the lifetime of a solution of fluorophores corresponds to the time for the intensity (following excitation by a narrow pulse) to decay to $1/e$, or 36.8%, of the beginning intensity. This point is illustrated in **Figure 7.4.** When quencher molecules, such as iodide, are introduced into the solution, the fluorophores most likely to collide with a quencher, and hence to be deactivated from the excited state, are the ones that "hang around" the longest. Hence, the longer-lived excited molecules will be preferentially removed from the excited state population, as depicted in **Figure 7.4.** The time for the intensity of this quenched population of excited fluorophores to decay to $1/e$ of the beginning intensity is thus shorter than in the absence of quencher.

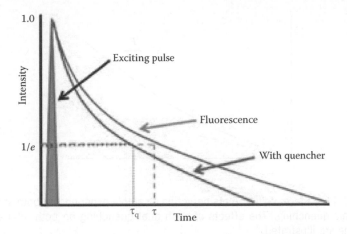

FIGURE 7.4 Sketch illustrating how quenching leads to shorter fluorescent lifetimes. The fluorescence decay curves are normalized to the same beginning intensity.

Static Quenching

In some cases, the fluorophore can form a stable complex with another molecule. If this *ground-state* is nonfluorescent, we say that the fluorophore has been statically quenched, a process depicted in **Figure 7.5**.

In such a case, the dependence of the fluorescence as a function of the quencher concentration follows the relation:

$$F_0/F = 1 + K_a[Q] \tag{7.7}$$

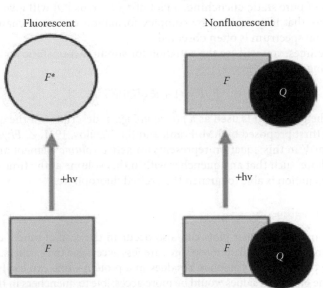

FIGURE 7.5 Depiction of fluorophore subject to static (ground-state) quenching.

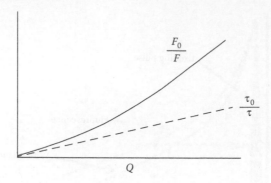

FIGURE 7.6 Stern–Volmer plots depicting a sample experiencing both static and dynamic quenching. The effects of such dual quenching on both intensity and lifetime are illustrated.

where K_a is the association constant of the complex. Such cases of quenching via complex formation were first described by Gregorio Weber, who showed that if both static and dynamic quenching occur in the sample, then the following relation holds:

$$F_0/F = (1 + k_q\, \tau[Q])(1 + K_a[Q]) \tag{7.8}$$

In such a case, a plot of F_0/F versus $[Q]$ will give an upward curving plot (**Figure 7.6**). The upward curvature occurs because of the $[Q]^2$ term in the equation. However, since the lifetime is unaffected by the presence of quencher in cases of pure static quenching, a plot of τ_0/τ versus $[Q]$ will give a straight line. Note that if quenching by complex formation occurs, a change in the absorption spectrum is often observed.

Sometimes, you will see the equation for simultaneous static and dynamic quenching given as

$$F_0/F = (1 + K_{SV}[Q])e^{V}[Q] \tag{7.9}$$

where the term $e^{V}[Q]$ is used as a *phenomological* descriptor of the quenching process (first proposed by I.M. Frank and S.I. Vavilov, 1931, *Z. Phys.* 69:100). The term V in this equation represents an *active volume* element around the fluorophore, such that any quencher within this volume at the time of fluorophore excitation is able to quench the excited fluorophore.

Other Considerations

Nonlinear Stern–Volmer plots can also occur in the case of purely collisional quenching if some of the fluorophores are less accessible than others. Consider the case of multiple tryptophan residues in a protein—one can easily imagine that some of these residues would be more accessible to quenchers in the solvent than others. In the extreme case, a Stern–Volmer plot for a system having accessible and inaccessible fluorophores could look like the plot shown in **Figure 7.7**.

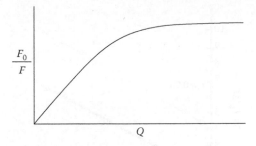

FIGURE 7.7 Illustration of quenching for the case of two fluorophores, one accessible to quencher and one inaccessible to quencher.

The quenching of LADH (liver alcohol dehydrogenase) intrinsic protein fluorescence by acrylamide gives, in fact, just such a plot in **Figure 7.8**. LADH is a dimer with two tryptophan residues per identical monomer. One residue is buried in the protein interior and is relatively inaccessible to acrylamide, while the other tryptophan residue is on the protein's surface and is much more accessible. In this case, the different emission wavelengths preferentially weigh the buried (323 nm) or solvent exposed (350 nm) tryptophans.

Sam Lehrer modified the Stern–Volmer equation to take into account situations wherein two populations of fluorophores exist—one that is accessible (a) to quencher, and the other that is buried and is not accessible (b). The

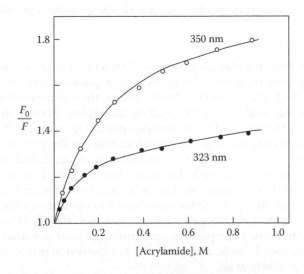

FIGURE 7.8 Illustration of quenching of LADH tryptophan fluorescence by acrylamide: fluorescence monitored at 350 nm or at 323 nm.

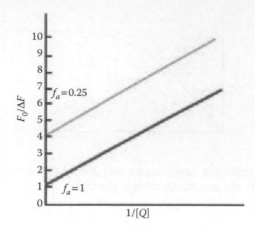

FIGURE 7.9 Illustration of Lehrer plot for fluorescence quenching for the cases of 100% accessible fluorescence (red) and 25% accessible fluorescence (blue).

observed fluorescence (F_{obs}) is then the fractional fluorescence due to a, or f_a. Rearrangement of the Stern–Volmer equation leads to

$$F_o/\Delta F = 1/([Q] f_a K_{SV}) + 1/f_a \qquad (7.10)$$

where ΔF is the observed decrease in the fluorescence and K_{SV} is the Stern–Volmer quenching constant of the accessible tryptophans. A plot of $F_o/\Delta F$ versus $1/[Q]$ (as shown in **Figure 7.9**), has a Y-axis intercept of $1/f_a$. In his original work of iodide quenching of lysozyme (which has multiple tryptophan residues), Lehrer showed that for lysozyme in pH 7.5 buffer, f_a was 0.38 whereas this value increased to unity for lysozyme in 6 M guanidinium chloride.

Popular Quenchers

One of the most popular quenchers used in protein studies is acrylamide. Pioneering work on acrylamide quenching of proteins was carried out by Maurice Eftink and Camillo Ghiron. Eftink then continued acrylamide studies and his work is essential reading for anyone wanting to enter this field. One of the reasons for acrylamide's popularity is that it is so easy to use. One need only add it to the cuvette containing the protein. A precaution, though, is to excite at 295 nm or higher since acrylamide will absorb increasingly as the wavelength decreases—hence, there is some difficulty if one wishes to study tyrosine residues. Also, acrylamide is far from a democratic quencher; that is, it will not quench all fluorophores equally. As shown by Eftink (Eftink et al., 1987, *Photochem. Photobiol.* 46: 23), although excellent for indole-related compounds, acrylamide is a poor quencher of various naphthalene-based compounds and will not quench acridine, fluorescein, eosin Y, or proflavin at all.

Iodide ion (I^-) is also a popular quencher because of its ease of use. Iodide has the drawback, however, that it is charged and hence its accessibility to a fluorophore will depend greatly on the electrostatic environment. In general, iodide ions can usually quench surface exposed tryptophan residues (depending on the charge environment), but do not effectively penetrate into protein interiors. Hence, iodide quenching can be useful for estimating the extent of tryptophan exposure on a protein's surface. One precaution required when using iodide to quench a macromolecule's fluorescence is to run control experiments with, for example, KCl instead of KI, to demonstrate (hopefully) that ionic strength alone does not affect the fluorescence. Another problem with iodide is the formation of I_3^- which absorbs in the ultraviolet. For this reason, small amounts (e.g., 0.1 mM) of sodium thiosulfate are usually added to eliminate any I_3^-.

Quenching of lifetimes of a few nanoseconds, for example, protein fluorescence, with molecular oxygen requires a pressure cell able to hold about up to 100 atm of oxygen. The reason for this complication is the relatively low solubility of molecular oxygen in aqueous solvents, for example, at 1 atm and 25°C the concentration of oxygen in an air-equilibrated solution is 0.001275 M. Hence, 100 atm of oxygen gas pressure will give about 0.13 M oxygen in aqueous solutions. On the other hand, if one happens to have an appropriate pressure cell, then oxygen quenching can be a valuable method to study quenching without the complications of charge or polarity of the quencher molecule. Such experiments on oxygen quenching of intrinsic protein fluorescence, carried out by Lakowicz and Weber (see Additional Reading), demonstrated that O_2 molecules were able to readily penetrate into protein interiors, and helped to establish the field of protein dynamics.

Quenching and Membrane Systems

Fluorescence quenching has also been used to study membrane systems, both model bilayers and natural membranes. NBD-labeled lipids, for example, have been widely used as fluorescent lipid analogues for quenching studies (these and other lipophilic probes are discussed in Chapter 10). In such studies, of course, the quencher must also be lipid soluble. The use of fluorophores and quenchers set at variable depths within a bilayer is now a well-established stratagem. This approach is based on the original studies of Keith Thulborn and William Sawyer (1977, *Biochemistry* 16: 982) who described the use of 9-anthrolyoxy fatty acid probes, with the fluorophore located at different positions along the fatty acid chain (**Figure 7.10**). Amitabha Chattopadhyay and Erwin London (*Biochemistry* 1987, 26: 39) described an approach to membrane quenching they called the parallax method. The method involves determination of the parallax in the apparent location of fluorophores, detected by comparing the quenching by phospholipids spin labeled at two different depths. By use of straightforward algebraic expressions, this method allows calculation of depth of the fluorophore within the membrane. In addition to quenching membrane-associated fluorophores, the tryptophan fluorescence of membrane proteins has also been extensively studied using quenching approaches. Another approach to membrane quenching is described by

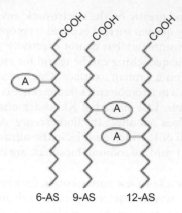

6-AS   9-AS   12-AS

FIGURE 7.10 Illustration of placement of fluorophores at different positions along a fatty acid backbone. In this figure, a series of n-(9-anthroyloxy) fatty acids are depicted. (Modified from Blatt and Sawyer, 1985. *Biochim. Biophys. Acta* 822: 43.)

Kyrychenko, Tobias and Ladokhin (*J. Phys. Chem. B* 2013, 117: 4770) who utilized a distribution analysis (DA) method of extracting quantitative information on membrane penetration from depth dependent fluorescence quenching experiments. The validity of their approach was established using Molecular Dynamics simulations.

Several types of quenching assays have been developed for membrane fusion. For example, in the case of vesicle fusion, one can label one population of vesicles with a fluorophore and another population with a quencher, as illustrated in **Figure 7.11** (top). Alternatively, one can have an overabundance of fluorophores in one vesicle population, sufficiently concentrated such that self-quenching occurs, for example, carboxyfluorescein at >100 mM, and then upon fusion of these vesicles with unlabeled vesicles, the dilution of the fluorophore concentration relieves the self-quenching, as illustrated in **Figure 7.11** (bottom).

As mentioned earlier, other types of quenching mechanisms, in addition to collisional quenching and complex formation, can occur. For example, energy transfer can also lead to depopulation of the excited state. The application of quenching via FRET has been widely applied to the area of biosensors. FRET processes in general will be covered in Chapter 8.

Fluorophores

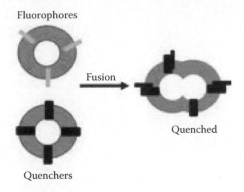

Quenchers

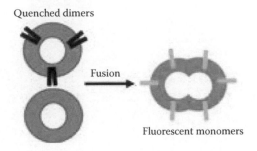

FIGURE 7.11 Depiction of two quenching approaches for fluorophores associated with vesicles. (Top) Vesicles containing fluorophores are mixed with vesicles containing quenchers; after vesicle fusion, quenching occurs. (Bottom) Vesicles containing fluorophores quenched by complex formation are mixed with vesicles with no fluorophores. Upon fusion, the fluorophore concentration per vesicle is reduced, which relieves the quenching.

Additional Reading

P.G. Seybold, M. Gouterman, and J. Callis, 1969. Calorimetric, photometric and lifetime determinations of fluorescence yields of fluorescein dyes. *Photochem Photobiol.* 9: 229–242.

J.R. Lakowicz and G. Weber, 1973. Quenching of fluorescence by oxygen. A probe for structural fluctuations in macromolecules. *Biochemistry* 12: 4161–4170.

M.R. Eftink, 1991. Fluorescence quenching: Theory and application in *topics in fluorescence spectroscopy*, Vol. 2, pp. 53–126. Ed. J.R. Lakowicz, Plenum Press, New York.

K. Rurack, 2008. Fluorescence quantum yields—Methods of determinations and standards in standardization and quality assurance. In *Fluorescence Measurements I. Techniques*, Vol. 5, pp. 101–145. Eds. U. Resch-Genger and O.S. Wolfbeis. Springer Series on Fluorescence, Springer-Verlag, Berlin.

Förster Resonance Energy Transfer

Background

What exactly is FRET and why has it become so popular in the past decade or two? Perhaps before starting this discussion, I should note that the Merriam-Webster online dictionary defines "FRET" as: "to cause to suffer emotional strain." Now I've warned you!

Simply put, FRET entails excitation of one fluorophore (in FRET parlance called the donor), followed by transfer of that excitation energy to another molecule (called the acceptor). The reader should note that I specifically did not state that the molecule receiving the energy, that is, the acceptor, was a fluorophore. In many applications the acceptor is, in fact, a fluorophore, which usually, but not always, is different from the donor fluorophore. In some cases, termed self-transfer or homoFRET, the acceptor fluorophore is the same as the donor. For example, one of the truly seminal papers in the field was by Enrique Gaviola and Peter Pringsheim in 1924, in which they observed fluorescein to fluorescein energy transfer. In other cases, the acceptor does not fluoresce at all (sometimes called a dark acceptor). As we shall see, the details of the spectroscopic properties of the donor and acceptor molecules will determine the appropriate techniques which can be used to detect and quantify the FRET process.

Why is FRET presently so popular in the physical and life sciences? No doubt because of the great many processes and reactions it allows one to follow. Some of the most popular uses of FRET in the life sciences were compiled by Bernard Valeur in his excellent book, *Molecular Fluorescence*. These include the use of FRET to study:

Ligand–receptor interactions

Conformational changes of biomolecules

Proteins: *In vivo* protein–protein interactions, protein folding kinetics, protein subunit exchange, enzyme activity assay, and so on

Membranes and models: membrane organization (e.g., membrane domains, lipid distribution, peptide association, lipid order in vesicles, membrane fusion assays, and so on)

Nucleic acid structures and sequences: primary and secondary structure of DNA fragments, translocation of genes between two chromosomes, detection of nucleic acid hybridization, formation of hairpin structure, interaction with drugs, DNA triple helix, DNA–protein interaction, automated DNA sequencing, and so on

Nucleic acid–protein interactions

Immunoassays

Biosensors

I should add that these FRET studies may be carried out *in vitro, in silico* or in living cells. In virtually all cases, the basic information being sought is whether or not two molecules are close to each other and if that distance changes as a result of some process. Ideally, one may hope to ascertain precisely the distance between donor and acceptor. In a well-known article, Lubert Stryer and Richard Haugland coined the term "spectroscopic ruler" for FRET (*Proc. Natl. Acad. Sci. USA*, 1967, 58, 719). As we shall learn, there are many caveats which limit the ability of FRET to determine distances accurately—but these difficulties do not mean that the method is not useful for a great many applications.

Radiationless energy transfer was described as early as 1922 when G. Cario and J. Franck (*Z. Phys.* 10: 185) noted that irradiation of a mixture of mercury and thallium atoms, in the gas phase, with a wavelength only absorbed by mercury, resulted in emission from both mercury and thallium. This observation indicated that transfer of energy could occur via a nonradiative process. *A brief comment here: by nonradiative transfer we mean that the transfer of energy is not simply due to one species emitting radiation, subsequently absorbed by another species.* Another important milestone in energy transfer, already mentioned, was the 1924 report by Enrique Gaviola and Peter Pringsheim that an increase in the concentration of uranin (sodium fluorescein) in a viscous solution (glycerol) was associated with a decrease in the polarization of the emission. These researchers realized that their observations suggested that some type of interaction could take place between fluorescein molecules that came near each other, that is, that a transfer of excited state energy could occur that did not require actual contact between the fluorescein molecules. Physicists including Jean and Francis Perrin, Fritz London, and even J. Robert Oppenheimer, worked on this problem and contributed to the theory, which culminated in the current day formulation given by Theodore Förster in 1946 (excellent histories of resonance energy transfer have been written by Robert Clegg). The phenomenon of radiationless energy transfer is now known as FRET, which stands for Förster resonance energy transfer. *Anecdote Alert!*

Interestingly, the switch from the general use of "Fluorescence Resonance Energy Transfer" to "Förster Resonance Energy Transfer" dates to the early 1990s at a Biophysical Society meeting during which Bob Dale, a well-known fluorescence practitioner, pointed out that using the word "fluorescence" for the letter "F" in FRET was misleading, since the act of energy transfer is in lieu of the fluorescence process. Bob suggested that the fluorescence community use the term "Förster Resonance Energy Transfer," which honors Förster's contributions, but which also preserves the letter "F" in FRET, which by that time was well entrenched in the literature.

To help grasp the fundamentals of the FRET phenomenon, consider a pair of tuning forks (perhaps you have seen this demonstration in primary school). If one strikes a tuning fork, then brings another tuning fork possessing the same fundamental vibration characteristics nearby, one will observe that the second tuning fork begins to vibrate. In fact, some of the vibrational energy of the first tuning fork is transferred to the second one. In this example, the first tuning fork could be termed the donor and the second one the acceptor. Similarly, if a fluorophore, the donor, is excited and another molecule (the acceptor) with the appropriate "vibrational" characteristics is nearby, it may be able to take on energy from the donor. Of course, the tuning fork analogy fails in many respects, but perhaps it conveys an impression of the phenomenon of energy transfer.

How is FRET most often implemented? A perusal of the literature indicates that three approaches dominate present day uses of FRET. In these cases, the donor molecule is fluorescent and the different cases arise from the nature of the acceptor molecule. These cases, illustrated in **Figure 8.1**, correspond to (1) donor and acceptor molecules differ and the acceptor is fluorescent, (2) donor and acceptor differ and the acceptor is not fluorescent, and (3) donor and acceptor are the same type of molecule.

Basic Principles

Precisely what properties of donor and acceptor molecules are important in FRET and how can we quantify the process? Let us begin by considering a simplified energy level diagram, similar to the Perrin–Jabłoński diagram we saw in Chapter 3 (**Figure 8.2**).

Suppose that the energy difference for one of the possible deactivation processes in the donor molecule matches that for a possible absorption transition in a nearby acceptor molecule. Then, with sufficient energetic coupling between these molecules (overlap of the emission spectrum of the donor and absorption spectrum of the acceptor), both processes may occur simultaneously, resulting in a transfer of excitation from the donor to the acceptor molecule. The interaction energy is of a dipole–dipole nature (**Figure 8.3**), and thus, depends on the distance between the dipoles and their relative orientation.

The rate of transfer (k_T) of excitation energy is given by

$$k_T = (1/\tau_d)(R_0/r)^6 \tag{8.1}$$

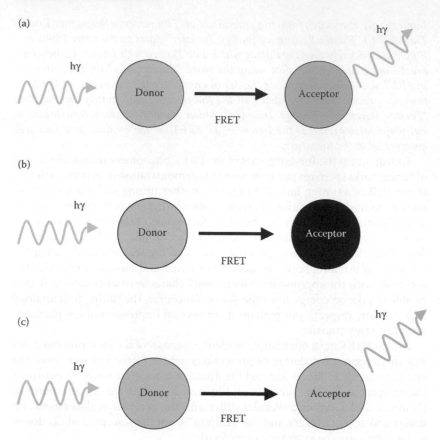

FIGURE 8.1 Illustration of three FRET approaches. (a) Donor and acceptor molecules differ and the acceptor is fluorescent, (b) donor and acceptor differ and the acceptor is not fluorescent, and (c) donor and acceptor are the same type of molecule.

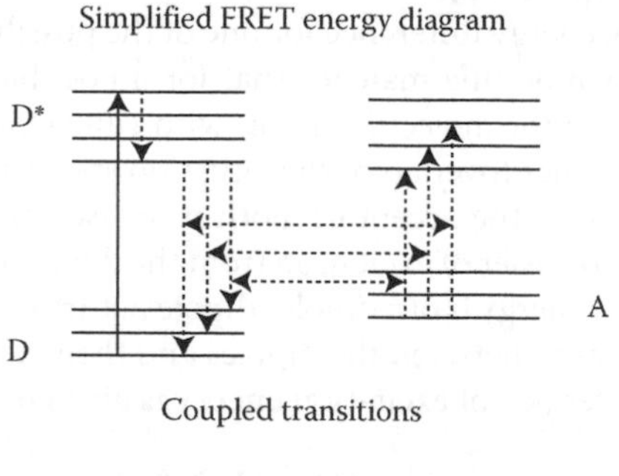

Simplified FRET energy diagram

Coupled transitions

FIGURE 8.2 Simplified energy level diagram illustrating FRET.

FIGURE 8.3 Donor and acceptor dipoles.

where τ_d is the fluorescence lifetime of the donor in the absence of acceptor, r is the distance between the centers of the donor and acceptor molecules, and R_0 is the Förster critical distance at which 50% of the excitation energy is transferred to the acceptor. Importantly, R_0 can be approximated from experiments independent of energy transfer.

Specifically,

$$R_0 = 0.2108(n^{-4}Q_d\kappa^2 J)^{\frac{1}{6}} \text{ Å} \qquad (8.2)$$

where n is the refractive index of the medium in the wavelength range where spectral overlap is significant (usually between 1.2 and 1.4 for biological samples), Q_d is the fluorescence quantum yield of the donor in the absence of the acceptor (i.e., number of quanta emitted/number of quanta absorbed), and κ^2 (pronounced "kappa-squared") is an orientation factor for the dipole–dipole interaction. J is the normalized spectral overlap integral:

$$J = \int_0^\infty I_D(\lambda)\varepsilon_A(\lambda)\lambda^4 \, d\lambda \qquad (8.3)$$

where $\varepsilon(\lambda)$ is in $M^{-1}\,cm^{-1}$, λ is the wavelength of the light in nm, $\varepsilon_A(\lambda)$ is the molar absorption coefficient at that wavelength and $I_D(\lambda)$ is the fluorescence spectrum of the donor normalized on the wavelength scale. The units of J is in $M^{-1}\,cm^{-1}\,nm^4$.

$I_D(\lambda)$ is given by

$$I_D(\lambda) = \frac{F_{D\lambda}(\lambda)}{\int_0^\infty F_{D\lambda}(\lambda)d\lambda} \qquad (8.4)$$

where $F_{D\lambda}(\lambda)$ is the donor fluorescence per unit wavelength interval. The overlap integral is illustrated in **Figure 8.4**.

FRET calculators, to calculate critical transfer distances, and so on, are readily available on the web, for example:

http://www.calctool.org/CALC/chem/photochemistry/fret

http://www.microscopyu.com/tutorials/java/fluorescence/fpfret/

Also, a great many R_0 values are listed in the literature for literally hundreds of donor–acceptor pairs.

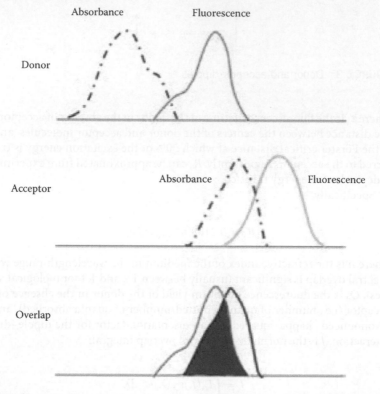

FIGURE 8.4 Depiction of overlap integral between donor emission and acceptor absorption.

How Do We Determine the Efficiency of Energy Transfer (*E*)?

The FRET efficiency, *E*, is the parameter most people want to determine, regardless of the application. It can be estimated in different ways, some of which are discussed below.

Steady-State Intensity

Perhaps, the most common way to estimate the efficiency of energy transfer is by determining the intensity of the donor emission in the absence (F_d) and presence (F_{da}) of an acceptor, as indicated in Equation 8.5. This process is illustrated in **Figure 8.5**.

$$E = 1 - \frac{F_{da}}{F_d} \tag{8.5}$$

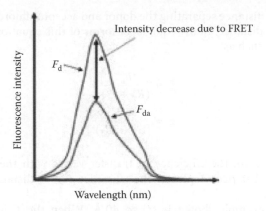

FIGURE 8.5 Depiction of the decrease in fluorescence intensity of the donor subsequent to FRET. F_d indicates the spectrum of donor alone while F_{da} indicates the spectrum of donor in the presence of acceptor.

Time-Resolved Method: Decrease in the Lifetime of the Donor

If the fluorescence decay of the donor is a single exponential, then

$$E = 1 - \frac{\tau_{da}}{\tau_d} \tag{8.6}$$

where τ_{da} and τ_d are the lifetimes of the donor in the presence and absence of an acceptor, respectively. If the donor fluorescence decay in the absence of an acceptor is not a single exponential (probably resulting from heterogeneity of the probe's microenvironment), then it may be modeled as a sum of the exponential and the transfer efficiency can be calculated using the average decay times of the donor in the absence and presence of the acceptor:

$$E = 1 - \frac{\langle \tau_{da} \rangle}{\langle \tau_d \rangle} \tag{8.7}$$

where $\langle \tau \rangle$ is the amplitude-average decay time and is defined as

$$\langle \tau \rangle = \frac{\sum_i \alpha_i \tau_i}{\sum_i \alpha_i} \tag{8.8}$$

The distance dependence of the energy transfer efficiency (E) is

$$r = \left(\frac{1}{E} - 1 \right)^{\frac{1}{6}} R_0 \tag{8.9}$$

where r is the distance separating the donor and acceptor fluorophores and R_0 is the Förster distance. Many equivalent forms of this equation are found in the literature, such as

$$E = \frac{R_0^6}{(R_0^6 + r^6)}$$

$$E = \frac{1}{[1 + (r/R_0)^6]} \tag{8.10}$$

As can be seen, the efficiency of transfer varies with the inverse sixth power of the distance. A plot of the efficiency versus distance is depicted in **Figure 8.6**.

R_0 in the example shown is set to 40 Å. When the E is 50%, $R = R_0$. Distances can usually be measured between ~0.5 R_0 and ~1.5 R_0. Beyond these limits, we can usually only assert that the distance is smaller than 0.5 R_0 or greater than 1.5 R_0. I should also point out that at short distances, on the order of 10 Å, another quenching mechanism, known as Dexter energy transfer (after David L. Dexter), may occur. Since Dexter transfer is not a consideration in most of the systems we are concerned with, I shall not discuss it in more detail.

Of course, assigning distances on the basis of the FRET efficiency and calculated R_0 assumes that the orientation factor is known, since, as shown in Equation 8.2, κ^2 is required for the R_0 calculation. Hence, anyone who is serious about assigning distances based on FRET, must be serious about κ^2.

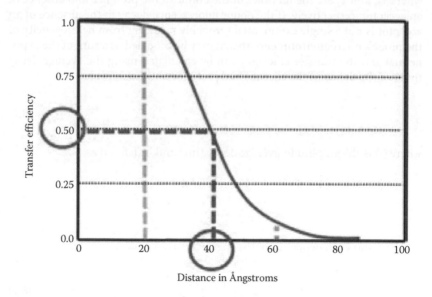

FIGURE 8.6 A plot of efficiency versus distance in a FRET system. R_0 is set to 40 Å.

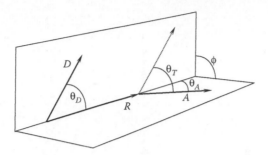

FIGURE 8.7 Depiction of the angles associated with the relative orientations of donor and acceptor dipoles.

The Orientation Factor

The orientation factor, κ^2, is related to the angle—in three-dimensional space—between the donor and acceptor dipoles, specifically,

$$\kappa^2 = (\cos\theta_T - 3\cos\theta_D\cos\theta_A)^2 \qquad (8.11)$$

where θ_T is the angle between the D and A moments, given by

$$\cos\theta_T = \sin\theta_D \sin\theta_A \cos\phi + \cos\theta_D \cos\theta_A \qquad (8.12)$$

In which θ_D and θ_A are the angles between the separation vector R and the D and A moment, respectively, and ϕ is the azimuth between the planes (D, R) and (A, R) (**Figure 8.7**). The limits for κ^2 are 0–4. The value of 4 is only obtained when both transition moments are in line with the vector R. The value of 0 can be achieved in different ways, as illustrated in **Figure 8.8**.

If the molecules undergo fast isotropic motions (dynamic averaging), then $\kappa^2 = 2/3$. When does this condition hold? It holds when the lifetime of the fluorophore is long, compared to the rotational relaxation time. If both donor and acceptor are fluorescent, and if the polarizations observed for both are near zero, then the assumption of fast isotropic averaging is probably reasonable.

How Do We Determine κ^2?

Except in rare cases, κ^2 cannot be uniquely determined in solution. Then, what value of κ^2 should be used? We can *assume* fast isotropic motions of the probes and set $\kappa^2 = 2/3$, and then try to justify, that is, verify experimentally, our assumption. For example, we can:

1. *Try swapping probes:* The micro-environment of the probes will almost certainly be different. Therefore, if the micro-environment affects the probes' mobility and κ^2 is not equal to 2/3, once swapped, the value of κ^2 will change and, hence, the FRET efficiency and the calculated distance. For example, in the case of a protein, if one is studying FRET between a tryptophan residue and an acceptor

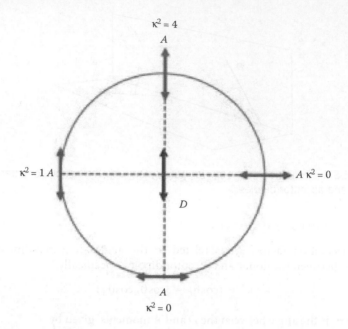

FIGURE 8.8 Illustration of the donor–acceptor orientations which lead to κ^2 values between 0 and 4.

linked to a cysteine residue, one may be able to carry out site-directed mutagenesis to swap the positions of the tryptophan and cysteine residues in the protein structure (**Figure 8.9**). Similar scenarios can be envisaged for different systems.

2. *Using different probes:* In many cases, one is able to use different probes for both donor and acceptor. If the FRET efficiencies, and hence distances, measured using different probe pairs are similar (taking into account the size of the probes), then the assumption that κ^2 is equal to 2/3 is probably valid, since the orientation of the probes' dipoles on their respective nuclear frameworks probably differ.

In some cases, one can estimate the lower and upper limit of κ^2 using polarization spectroscopy, as pointed out by Robert Dale, Josef Eisinger, and

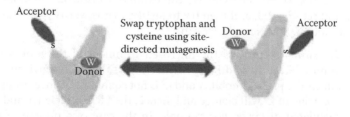

FIGURE 8.9 Illustration of the strategy of swapping donor and acceptor positions.

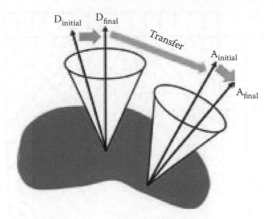

FIGURE 8.10 Illustration of mobilities of both donor and acceptor dipoles, which lead to depolarization.

William Blumberg in 1979 (*Biophys. J.* 26: 161). Let us consider that each probe, donor and acceptor, is free to rotate within a cone. Then, three depolarization steps occur after the absorption of the excitation energy by the donor: an axial depolarization of the donor, a depolarization due to transfer and an axial depolarization of the acceptor (**Figure 8.10**).

In the Dale–Eisinger–Blumberg approach, one measures the ratio of the observed anisotropies of donors and acceptors to their limiting anisotropies, and then contour plots, such as that shown in **Figure 8.11**, are used to put limits on κ^2. For example, one can see from this plot that for cases wherein the polarizations/anisotropies of both donor and acceptor are near zero, κ^2 will be near 2/3. If both donor and acceptor polarizations/anisotropies are near their limits, however, then κ^2 can be between values higher than 3 and lower than 0.1 (see, e.g., the upper right corner of **Figure 8.11**). These determinations are actually a bit more complicated than indicated, and readers with a sustaining interest are referred to the original paper. What is clear, however, is that systems wherein the donor and acceptor have short lifetimes, relative to their rotational motions, will have high polarizations and hence increased uncertainties in the correct κ^2 value. Certainly, fluorescent proteins (e.g., GFP, CFP, YFP, etc.), especially when they are linked to other proteins, fall into this category.

Quantitative distance determinations using FRET, that is, as a true "spectroscopic ruler," remain difficult at best. But FRET can be very powerful when used to detect *changes* in a system, such as alterations in distance and/or orientation between donor and acceptor attached to biomolecules, that is, due to ligand binding or protein–protein interactions.

Homo-Transfer of Electronic Excitation Energy

So far, we have considered the donor and acceptor molecules to be different. However, if the probe excitation spectrum overlaps its emission spectrum,

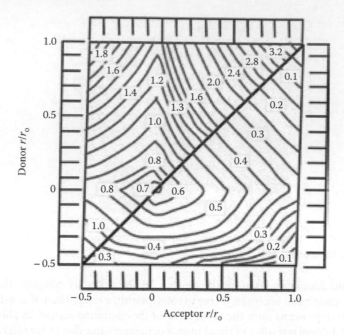

FIGURE 8.11 A Dale–Eisinger–Blumberg contour plot, which put limits on the value of κ^2.

FRET can occur between identical molecules. Of course, if the donor and acceptor are identical molecules, one cannot use spectral differences between donor and acceptor to study the FRET process. However, such homoFRET can lead to a change in polarization. In fact, Francis Perrin recognized that excited state energy transfer could lead to depolarization of the emission when he wrote (F. Perrin, *Ann de Phys.* 1929):

> Il suffit qu'un transfert d'activation puisse se produire entre deux molécules voisines d'orientation différentes, c'est a dire portant des oscillateurs non parallèles, pour qu'il en résulte en moyenne une diminution de l'anisotropie de distribution des oscillateurs excites et par suite de la polarisation de la lumière émise.

Which can be translated as:

> It suffices that a transfer of activation can occur between two neighboring molecules with different orientations, that is with non-parallel oscillators, in order to have, on average, a decrease in the anisotropy of the distribution of excited oscillators, and therefore a decrease of the polarization of the emitted light.

The phenomenon of depolarization via energy transfer was discussed in Chapter 5, and illustrated in Figure 5.29. Electronic energy transfer between identical fluorophores was originally observed by Gaviola and Pringsheim in 1924, as discussed in Chapter 5. Hamman et al. (1996, *Biochemistry* 35: 16680) used fluorescein homoFRET to study the dimer to monomer dissociation of the eubacterial ribosomal protein L7/L12. One way they verified the homo-FRET process was to observe the failure of energy transfer upon excitation of the fluorescein near the red-edge of its absorption. This phenomenon is called Weber's Red Edge effect after Gregorio Weber, who first observed it in 1960 in concentrated solutions of indole (as mentioned in Chapter 11). In 1970, Weber and Shinitzky published a more detailed examination of this phenomenon. They reported that in the many aromatic molecules examined (such as fluorescein), transfer was much decreased or undetectable upon excitation at the red edge of the absorption spectrum (Weber and Shinitzky, 1970, *Proc. Natl. Acad. Sci. USA* 65: 823).

Examples of FRET Applications

FRET studies are prevalent in the chemical and life sciences and examples of such studies are mentioned throughout this book. Here I shall give only a few specific examples to demonstrate some additional applications.

In Vitro

FRET is often used for diverse assays, both *in vitro* and in cells. Protease assays based on FRET are fairly common. The general strategy for such assays is illustrated in **Figure 8.12**. Fluorescence parameters which can be used to monitor such assays include intensities, lifetimes, and polarizations. As an example of an *in vitro* FRET-based assay, I shall mention a system I have worked on, namely botulinum neurotoxin. Botulinum neurotoxin A contains a zinc protease, which cleaves the protein, SNAP-25, which is involved in docking of synaptic vesicles with the presynaptic membrane. To detect active toxin, Gilmore et al., (2011, *Anal. Biochem.* 413: 36) utilized a biosensor comprising blue fluorescent protein (BFP) linked via residues from SNAP-25 to green fluorescent protein (GFP). This assay, termed DARET (depolarization after resonance energy transfer) was carried out by exciting into the BFP absorption band (at 387 nm) with polarized excitation and then observing the polarization of the emission from the GFP moiety (at 509 nm). In the intact substrate, FRET occurs between the BFP and GFP moieties, which results in depolarization of the emission (since the BFP and GFP dipoles are oriented differently). Upon cleavage of the substrate, and separation of the BFP and GFP moieties, the emission observed at 509 nm is only from direct GFP excitation (with the polarized excitation), and the polarization of the emission is high (**Figure 8.13**). This increase in polarization, from below zero to greater than 0.3, allows for detection and quantification of the active toxin (note that the starting negative polarization is possible because of the FRET process). By using the additivity function of polarization/anisotropy (discussed in Chapter 5), it is possible to assess the extent of substrate cleavage, which permits calculation of the enzymatic parameters, such as the Michaelis–Menton constant. The photophysics underlying the DARET assay were discussed in Ross et al. (2011 *Anal. Biochem.* 413: 43).

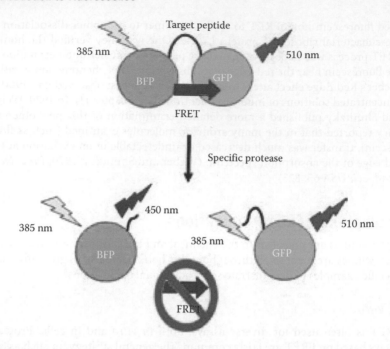

Target peptide

385 nm   510 nm

BFP   GFP

FRET

Specific protease

450 nm

385 nm   510 nm

385 nm

BFP   GFP

FRET

FIGURE 8.12 Depiction of a FRET-based protease assay. Excitation of the BFP moiety results in significant energy transfer to GFP, which emits near 510 nm. Upon protease treatment and subsequent separation of the BFP and GFP proteins, excitation near 385 nm results in both BFP and GFP emission.

Energy transfer can be observed with intrinsic protein fluorescence. As early as 1960, Gregorio Weber showed that electronic energy transfer could take place from tyrosine to tryptophan (heterotransfer) as well as between tryptophan residues (homotransfer). Examples of such protein-related studies will be discussed in Chapter 11.

In Cells

The development of fluorescent proteins has led to a significant increase in FRET studies in cells. Some examples of these studies are given in Chapter 10, for example, the Cameleon system for intracellular calcium determinations. Other biosensors utilized in cells are the FLARE systems, also discussed in Chapter 10. Different approaches for the detection and quantification of the FRET phenomenon in cells (as well as *in vitro*) are used, for example:

1. *Intensity ratio methods*: In these approaches, the change in FRET efficiency is measured using the change in the ratio of the donor emission to the acceptor emission. An example is shown in Figure 10.45 for the Cameleon biosensor. A variation of the intensity ratio method is the technique of acceptor photobleaching. In this method, the ratio of donor to acceptor emissions is measured

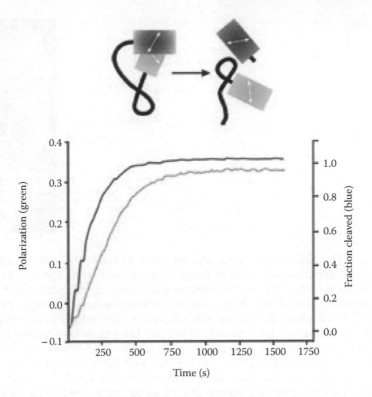

FIGURE 8.13 (Top) Depiction of the principle of the DARET assay. The BFP and GFP moieties are linked via residues from SNAP-25, the substrate for the Botulinum neurotoxin A zinc-protease. After cleavage, the two protein moieties can separate. (Bottom) Plot showing how the polarization from the GFP is largely depolarized before cleavage of the SNAP-25 peptide, and that as cleavage of the substrate progresses, the polarization of the GFP emission increases.

before and after intense illumination of the sample at a wavelength absorbed by the acceptor but not by the donor. The idea is to bleach, that is, effectively remove, the acceptor, which will then give rise to an increase in the donor emission—if a FRET process had been present before bleaching.

2. *Donor lifetime*: The advances in fluorescence lifetime imaging microscopy (FLIM) (discussed in Chapter 9) have led to an upsurge in the popularity of donor lifetime studies. As indicated by Equation 8.6, determination of the donor lifetime in the absence and presence of an acceptor allows determination of the FRET efficiency. An example of the use of FLIM to observe FRET in cells is shown in Figure 9.22. The phasor approach to lifetimes, mentioned in Chapter 6, has also gained popularity for FLIM–FRET studies. An illustration of the phasor plot approach to FLIM–FRET data is shown in **Figure 8.14**.

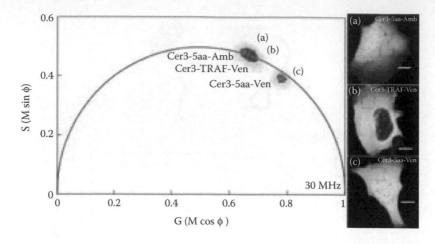

FIGURE 8.14 Illustration of the phasor approach to FLIM-FRET. (a) The phasor lifetime distribution and intensity image of a cell expressing Cerulean3–5aa-Amber (unquenched donor). The lifetime distribution on the phasor plot represents every pixel in the image, and the intensity-weighted average lifetime of 3.96 ns was determined in the ROI indicated by the red box in the intensity image; the calibration bar indicates 10 μm. (b) The phasor lifetime distribution and intensity image of a cell expressing Cerulean3-TRAF-Venus. The average lifetime of 3.68 ns was determined in the indicated ROI. (c) The phasor lifetime distribution and intensity image of a cell expressing Cerulean3–5aa-Venus. The amplitude-weighted average lifetime of 2.36 ns was determined in the indicated ROI. (The author would like to thank Richard N. Day for this figure from R.N. Day (2013) *Methods* http://dx.doi.org/10.1016/j.ymeth.2013.06.017.)

Advanced Treatments

FRET is commonly used in single molecule studies, as discussed in Chapter 9 (see, e.g., Figure 9.25). Recently, Claus Seidel and coworkers have described methods based on explicit considerations of spatial distributions of dye positions (Kalinin et al., 2012 *Nat. Methods* 9: 1218). Their approach, based on FRET-restrained high-precision structural modeling, has produced excellent agreement with x-ray-based structures in test systems. A discussion of their methods is beyond the scope of this introductory book, but readers with a sustaining interest in this topic are referred to their paper (listed below in Additional Reading).

Additional Reading

B.W. van der Meer, G. Coker Ill, and S.-Y. Chen, 1994. *Resonance Energy Transfer: Theory and Data*. VCH Publishers (now Wiley-VCH), Inc., New York.

B.D. Hamman, A.V. Oleinikov, G.G. Jokhadze, R.R. Traut, and D.M. Jameson, 1996. Dimer/monomer equilibrium and subunit exchange of *Escherichia coli* ribosomal protein L7/L12. *Biochemistry* 35:16680–16686.

R. D. Clegg, 2006. The history of FRET: From conception through the labors of birth. *Reviews in Fluorescence* 1–45.

V.V. Didenko (Ed.), 2006. *Fluorescent Energy Transfer Nucleic Acid Probes: Designs and Protocols.* Humana Press, Totowa, New Jersey.

S. Kalinin, T. Peulen, P.J. Rothwell, S. Berger, T. Restle, R.S. Goody, H. Gohlke, and C.A.M. Seidel, 2012. A toolkit and benchmark study for FRET-restrained high-precision structural modeling. *Nat. Methods* 9:1218.

Brief Overview of Fluorescence Microscopy

Up until this point, I have primarily discussed fluorescence measurements carried out *in vitro*. In fact, almost all of the measurements discussed so far can also be carried out in living cells using fluorescence microscopy. I suspect that many readers are interested in working with live cells so I will briefly discuss fluorescence microscopy. However, the focus of this discussion will be on quantitative aspects of modern fluorescence microscopy, in particular, its applications to biophysical issues associated with molecular cell biology. My treatment of this topic will be more of a survey of techniques and applications.

Fluorescence microscopes have actually been around for a long time. Otto Heimstaedt and Heinrich Lehmann developed the first fluorescence microscopes between 1911 and 1913, as an outgrowth of the UV microscope (August Köhler and Moritz von Rohr, 1904). These instruments were used to investigate the autofluorescence of bacteria, protozoa, plant and animal tissues, and bioorganic substances such as albumin, elastin, and keratin. Stanislav Von Prowazek (1914) employed the fluorescence microscope to study dye binding to living cells. During the past two decades, the use of fluorescence microscopes has grown tremendously—due not only to technical innovations (such as confocal optics and multiphoton lasers), and analysis methods (such as correlation approaches), but also due to the huge increase in fluorescent probes (including fluorescent proteins). Some of these innovations will be discussed in this chapter.

Introduction of Fluorophores into Living Cells

Unless one is interested in the intrinsic fluorescence associated with living cells, that is, autofluorescence (some of the molecules causing autofluorescence are discussed in Chapter 10), one must introduce fluorescent molecules into the cell. In some cases, this step can be accomplished via microinjection or by using fluorophores or profluorophores, which can be taken up directly by the cell. For example, esters of many xanthene dyes, such as fluorescein diacetate (**Figure 9.1**), are nonfluorescent and will passively diffuse across cell membrane (although in some cases active transport mechanisms have been proposed). Once inside the cell, endogenous esterases hydrolyze the esters to the highly fluorescent anion forms, which cannot diffuse back out of the cell. A wide range of calcium probes, such as Fura-2, Fluo-4, Calcium Green, and others have been used in this way to monitor calcium ion concentrations in cellular interiors. Some of these probes are discussed in Chapter 10. Fluorescent lipid analogs have been introduced into the plasma membrane of living cells by incubating the cells with lipid–BSA complexes. Other methods of introducing fluorophores into living cells include electroporation or the use of poreforming agents such as streptolysin-O. Fluorescently tagged antibodies are also widely used, especially to visualize antigens, such as receptors, on the cell surface. Fluorescent ligands can also be taken up by cells by receptor endocytosis.

The most popular methods nowadays for introducing fluorophores into cells involve genetic approaches. The original observation of green fluorescent protein (GFP) in the jellyfish *Aequorea victoria* was made by Osamu Shimomura in the early 1960s. Shimomura and his colleagues succeeded in isolating GFP in the 1970s, and Doug Prasher cloned the cDNA for GFP in 1992. Later in 1992, Martin Chalfie's laboratory expressed GFP in bacteria and in 1995 Tullio Pozzan's laboratory expressed GFP cDNA in cultured mammalian cells. Since that time, the application of recombinant fluorescent proteins (FPs) has exploded. This upsurge in popularity was largely due to the work by Roger Tsien and his colleagues on mutations in the protein sequence that altered the absorption and emission properties of GFP and allowed development of FPs which were ideal for observing FRET processes, and which were

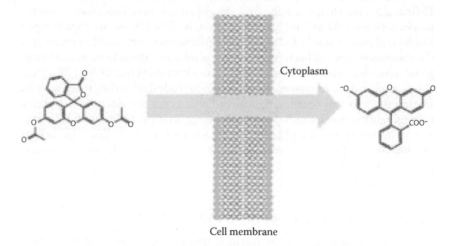

FIGURE 9.1 Illustration of the uptake of the nonfluorescent ester form of a fluorophore across a cell membrane, followed by hydrolysis to a highly fluorescent anion form, which cannot diffuse back out of the cell.

also more suitable for use in commercial microscopes. A more detailed discussion on FPs is presented in Chapter 10 along with other *in vivo* probes.

Fluorescence Microscopy Approaches

Many of the basic principles of a fluorescent microscope are similar to those of regular fluorimeters, namely, light of a particular wavelength is isolated from a light source and focused on a sample. The emitted light is then collected, isolated using a wavelength selection device to block excitation light, and then quantified using a photodetector. The essential component to route light, the filter cube, was described in Chapter 3. Obviously, one big difference between microscopy and cuvette studies is the spatial resolution microscopy provides. Readers with a sustaining interest in microscope optics should consult specialized texts (e.g., *Biological Confocal Microscopy* edited by James Pawley—listed under Additional Reading at the end of this chapter). The websites of microscope manufacturers, such as Nikon, Olympus, Leica, and Zeiss, among others, also provide a lot of information. Here, I shall give only the "bare bones" of fluorescence microscopy and its applications.

Confocal Microscopy

The most commonly used, and generally the least expensive, fluorescence microscopes are what are termed epifluorescence microscopes. An example of an epifluorescence microscope, in particular, what is termed an inverted microscope since the light enters the sample from below, is shown in **Figure 9.2**. A common light source is the mercury arc lamp, which, as mentioned in Chapter 3, provides light over a broad wavelength range but which has several very intense emission

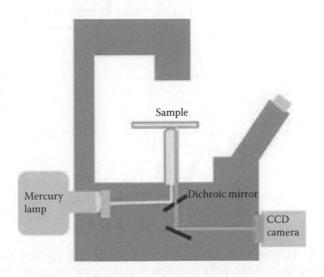

FIGURE 9.2 Depiction of an epifluorescence microscope.

wavelengths. Bandpass filters are used to select the wavelength region for sample illumination and a dichroic mirror (see Chapter 3) is used to reflect the light into the objective lens, which focuses it onto the sample. The fluorescence is then collected by the same objective lens and passes through the dichroic mirror to an additional bandpass filter (termed a barrier filter by microscopists), which removes any reflected excitation light and passes the selected wavelengths only. The fluorescence photons then impinge on the detector, typically a CCD (charge-coupled device) camera, although less expensive systems may simply use ocular detectors, that is, the eyes. These types of microscopes, although essential to many biological observations, do not generally allow the type of biophysical determinations I wish to discuss. Hence, I shall focus my attention on what are termed confocal microscopes. In normal epifluorescence, or widefield fluorescence microscopy, the fluorescence light reaching the detector does not arise from a unique focal plane. In fact, when the sample is thicker than the focal plane of the microscope objective, light originating from different depths within the sample can reach the detector. Confocal microscopy, however, takes advantage of the fact that emission photons coming from different depths within the sample—for example, a cell—are spatially separated as they traverse the microscope's detector optics. By placing an aperture, a small diameter pinhole, at the proper position in front of the detector, light not originating from the focal point is blocked, which results in sharper images than conventional optics. This principle is illustrated schematically in **Figure 9.3** and a comparison of a normal

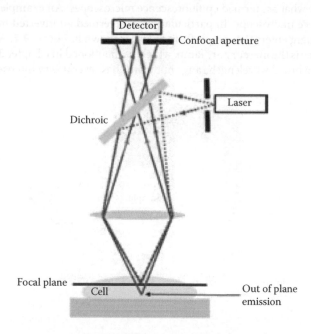

FIGURE 9.3 Illustration of the operational principle of a confocal microscope. The confocal aperture prevents emission originating outside of the focal plane from reaching the detector.

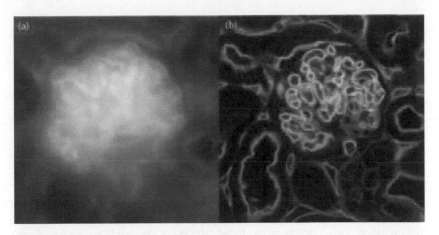

FIGURE 9.4 Comparison of wide-field (a) and confocal image (b). Sample is a 50 µM thick section of mouse kidney labeled with Isolectin B4 and Wheat Germ Agglutinin. (Reproduced with permission from L. Fritzky and D. Lagunoff, 2013. *Anal. Cell. Pathol.* 36: 5; original image by Luke Fritzky.)

wide-field image with the same image obtained using confocal optics is shown in **Figure** 9.4.

In addition to the inherent improvement in the image quality, another advantage of this approach is that the focal point can be moved up and down along the Z-axis, which permits one to obtain images of different sections within the sample. These sections can be assembled to give a three-dimensional reconstruction of the sample. Although confocal microscopy was first proposed in 1955 by Marvin Minsky, it became widely adopted only after the development of laser scanning techniques in the late 1970s. In the raster-scanning approach, the position of the laser light on the sample is controlled by a pair of galvanometric mirrors, which can be moved independently of one another, so as to move the exciting light in a repetitive, or raster, pattern, as depicted in **Figure** 9.5. An alternative approach to scanning involves movement of the sample stage under a fixed laser position. Each approach offers advantages and disadvantages, but detailed consideration of these issues is beyond the scope of this treatise. In either method, the individual pixels which form the image are illuminated typically for very short times. In the scanning mirror approach, the dwell time per pixel can be on the order of microseconds.

In addition to the raster scanning confocal approach, other confocal methods exist, such as the spinning disk microscope, which is essentially a multi-beam scanning technique. In the traditional spinning disk method, the exciting light passes through two mechanically spinning disks, also known as Nipkow disks after the inventor. The first implementation of a spinning disk confocal microscope was realized by David Egger and Mojmír Petráň in 1967. In the two disk arrangement, the first disk is typically fabricated with a number of equally spaced microlenses, while the second disk has a series of pinholes. In this approach, the sample is illuminated with multiple spots

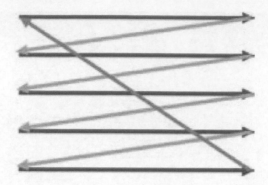

FIGURE 9.5 Depiction of a raster scanning process. The scan begins at the upper left and progresses to the right (blue arrow). When the end of the line is reached the excitation source rapidly moves to the left side of the next scan line (red arrow), then this process is repeated until the bottom right corner is reached at which point the excitation source moves rapidly to the beginning position (gray arrow).

simultaneously, essentially a parallel processing system, that allows for much faster frame acquisition rates (typically ~50 frames/s) compared to raster scanning confocal microscopy (typically on the order of a second or less per frame). The emission passes back through the second (pinhole) disk and is then diverted by a dichroic mirror to the detector. Single-sided spinning disk confocal microscopes, which utilize the same set of pinholes for both illumination and detection, were designed by Gordon Kino and Jeff Lichtman during the late 1980s. Modern spinning disk systems can have disks containing around 200,000 pinholes, with diameters of 25 μm, spinning at up to 2000 rpm, to give scanning rates of around 700 frames/s. The raster scanning and spinning disk approaches each have their advantages and disadvantages, but I shall not go into further detail. As always, readers with a sustaining interest in these topics are referred to the copious literature.

A more recent approach to confocal imaging is single plane illumination microscopy (SPIM), in which a thin sheet of light is used to illuminate (using cylindrical lenses) different planes in the sample at right angle to the observation direction. This "light sheet" can be very thin (<2 μm), and the entire image is typically collected using an array detector, such as a CCD. The plane illumination can be centered at various depths within the sample, which allows for sectioning and hence, three-dimensional reconstruction. I shall now consider some specialized fluorescence microscope techniques which have had, and continue to have, a significant impact in cell biology.

Total Internal Reflection Fluorescence

The TIRF method utilizes the fact that a beam of light propagating through a medium of one refractive index, impinging at an appropriate angle upon another medium of a smaller refractive index, will be totally reflected from the interface. However, some of the energy of the incident beam will penetrate

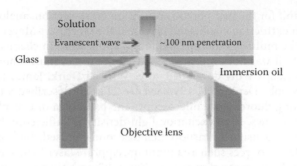

FIGURE 9.6 Illustration of the total internal reflection fluorescence (TIRF) method.

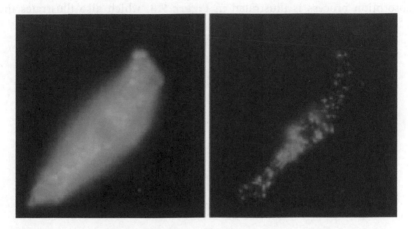

FIGURE 9.7 A comparison of an epifluorescence (left) and TIRF (right) image of a cell expressing GFP-marked chromaffin granules. (The author wishes to thank Daniel Axelrod for these images.)

through the interface, creating what is termed an evanescent field, which extends a very short distance, on the order of 100 nm, into a second medium. Hence, this evanescent beam will only be able to excite fluorophores which are located near the surface of the interface. This concept is illustrated in **Figure 9.6**. The TIRF method is hence ideally suited for studying processes which occur on or in the proximity of a cell's plasma membrane. A TIRF image is shown in **Figure 9.7**. Total internal reflection is also used in the technique known as surface plasmon resonance, which has become a popular method for studying molecular interactions *in vitro*.

Multi-Photon Excitation

Maria Goeppert-Mayer (who won the 1963 Nobel Prize in Physics for her work on the structure of nuclear shells) predicted the possibility of nonlinear, multi-photon absorption in her 1931 doctoral thesis. In honor of her work, the unit for the two-photon absorption cross-section, analogous to the one-photon extinction coefficient, is named the Goeppert-Mayer (GM) unit. Although the multi-photon method found application in chemistry as early as the 1960s, it was not until the 1990s that the use of multi-photon fluorescence microscopy truly developed, after Wilfred Denk, James H. Strickler, and Watt Webb (Denk et al. 1990 *Science* 248: 73) described a two-photon laser scanning fluorescence microscopy. The basic idea of two-photon excitation is as follows. If an excitation light density is sufficiently high, which can be achieved using ultrashort (femtosecond), pulsed, high peak power (kilowatt) laser sources such as a titanium:sapphire laser (which emits in the near-infrared, over a range of around 700–1000 nm), then two photons, each at approximately half the wavelength of the normal one-photon absorption of the target chromophore, can be simultaneously absorbed. The two-photon absorption process is illustrated in **Figure 9.8**, which also illustrates the

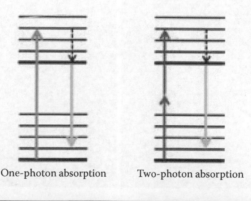

One-photon absorption   Two-photon absorption

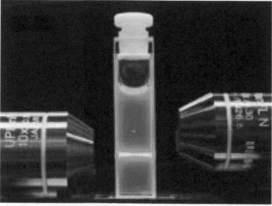

FIGURE 9.8 A comparison of the one-photon and two-photon absorption processes and an illustration of a fluorescein solution subjected to one-photon (380 nm; bottom) and two-photon (760 nm; top) illumination. (The author would like to thank Kevin Belfield for the fluorescein picture from http://chemistry.cos.ucf.edu/belfield/photophysics.)

intrinsic confocal aspect since the two-photon effect will be manifested only at the focal point of the lens. It is important to realize, however, that two-photon excitation follows different selection rules than one-photon excitation, and hence the shape and position of a fluorophore's two-photon absorption cross-section can differ somewhat from that expected from the one-photon absorption spectrum (**Figure 9.9**). A large effort is presently being expended on the development of fluorophores with improved two-photon absorption cross-sections.

Two-photon excitation, coupled with a fluorescence microscope, offers several advantages over traditional one-photon excitation, namely, (1) a very large separation between the excitation and emission wavelengths (which greatly minimizes problems associated with scattered light), (2) an intrinsic confocal aspect, that is, without the need for an aperture to limit light reaching the detector, and (3) reduced overall photobleaching, since light outside of the focal region is at a wavelength (infrared) not usually absorbed by the sample.

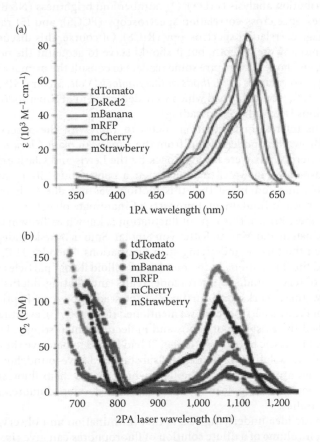

FIGURE 9.9 One- (a) and two-photon (b) absorption of a series of fluorescent proteins. (Modified from M. Drobizhev et al., 2011. *Nature Methods* 8: 393.)

171

Specialized Techniques

Fluorescence Fluctuation Spectroscopy (FFS)

It has been said that "One man's noise is another man's signal," and nowhere is this aphorism better illustrated than in the field of fluorescence fluctuation spectroscopy (FFS)—unless we consider also the area of rock music! Fluctuations in a fluorescence signal can arise from many causes, for example, diffusion of particles in and out of the observation volume, creation, and/or destruction of fluorescence molecules, conformational dynamics which alter quantum yields of fluorophores associated with macromolecules, intrinsic "blinking" processes of fluorophores, and so on. FFS methods allow us to study fluctuation processes and to extract useful information about the systems under study. I shall present a brief overview of some of the more popular FFS methods presently being applied to problems of cellular dynamics. The particular methods I shall discuss include: (1) fluorescence correlation spectroscopy (FCS); (2) photon counting histogram (PCH) and fluorescence intensity distribution analysis (FIDA); (3) number and brightness (N&B) analysis; (4) fluorescence cross-correlation spectroscopy (FCCS), and (5) raster scanning image correlation spectroscopy (RICS). Of course, this discussion cannot go into any great depth, but it should serve to acquaint the novice with the area, and hopefully inspire some readers to consult the primary literature. Recently, two volumes of *Methods in Enzymology* (Vols. 518 and 519, edited by Sergey Tetin), were published which were dedicated to FFS methodologies and applications (see Additional Reading).

Serious interest in fluctuations dates to 1827 with the observations by Robert Brown that pollen grains from the American plant *Clarkia pulchella*, (which interestingly were brought back by the Lewis and Clark expedition), when suspended in water, demonstrated a continuous, jittery movement. Brown became convinced that this motion was not due to living matter, but rather, was a fundamental property of inanimate objects in solution. To honor Robert Brown, this type of fluctuation is known as Brownian motion. Albert Einstein and Marian Ritter von Smolan Smoluchowski independently developed the theory underlying such fluctuations. Theodor H.E. Svedberg observed the fluctuations in the number of colloidal gold particles in a small volume observed under a microscope, really anticipating fluctuation spectroscopy. These works, and further elegant studies by Jean Baptiste Perrin (father of Francis Perrin whom we mentioned in Chapter 5), eventually firmly established the existence of atoms and molecules. Interestingly, Jean Perrin stated in his classic book, *Les Atomes*, "I had hoped to perceive these fluctuations in dilute solutions of fluorescent substances. I have found, however, that such bodies are destroyed by the light which makes them fluoresce." Thus, but for photobleaching, Jean Perrin might have developed fluorescence fluctuation studies!

The basic idea underlying FFS is that illumination and observation of a very small volume of a dilute solution of fluorophores can give rise to a fluorescence signal that displays considerable fluctuations due to the relatively small number of fluorescing molecules in the illumination/observation

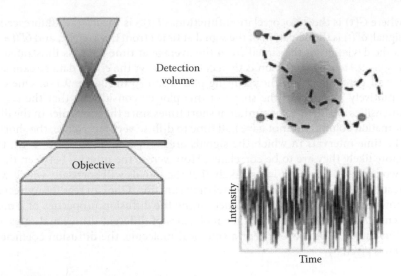

Detection volume

Objective

Intensity

Time

FIGURE 9.10 Illustration of the illumination of a small sample volume and detection of the emission from that volume. This volume corresponds to the point spread function (PSF). Also illustrated is the fluctuation in the fluorescence intensity as fluorophores diffuse into and out of the PSF. (From D.M. Jameson et al., 2009. *Biophys. Rev.* 1: 105.)

volume, as illustrated in **Figure 9.10**. In fact, significant information can be extracted from such a seemingly noisy signal. When we say "illumination and observation of a very small volume," how small are we talking about? That question brings us immediately to the concept of the *point spread function* or PSF. A PSF is illustrated by the red ellipse in **Figure 9.10**. The precise shape and dimension of a PSF will depend upon several factors, and determination of these parameters is a really complicated process best left to the experts. Let us just say that the typical PSF encompasses a volume on the order of a few femtoliters (which is 10^{-15} L). A typical PSF for two-photon excitation resembles an ellipsoid, around 0.3 μm in the XY directions (the plane perpendicular to the light direction) and 1.5 μm in the Z direction (along the direction of the exciting light). I note that typically FCS practitioners do not determine the PSF directly but rather, use a standard of known diffusional rate, such as fluorescein or rhodamine, to estimate the PSF. Recent careful determinations of diffusional rates of various xanthene-based dyes (fluorescein, rhodamine, Alexa) suggests that a value near 430 $\mu m^2 s^{-1}$ is reasonable in these cases.

One of the mathematical approaches used to extract information from the signal involves the *autocorrelation function*. This function is given by

$$G(\tau) = \frac{\delta F(t)\delta F(t + \tau)}{F(t)^2} \tag{9.1}$$

where $G(\tau)$ is the autocorrelation function, $<F(t)>$ is the average fluorescence signal, $\delta F(t)$ is the deviation of the signal at time t from the average, and $\delta F(t + \tau)$ is the deviation of the signal from the average at time $t + \tau$, as illustrated in **Figure 9.11a**. This function is then calculated over the entire data stream and for all τ intervals, and the resulting plot is shown in **Figure 9.11b**. One can intuitively understand the shape of this plot by considering that the signal intensities are likely to be similar at short times since the molecules in the illumination volume will not have had time to diffuse very far. Hence, the shorter the time intervals in which the signals are compared, that is, t to $t + \tau$, the more likely they are to be correlated. However, as the interval between these two points increases, it is less likely that the signals will correlate, which will result in a decrease in the autocorrelation function. One can readily appreciate how this method allows one to determine the diffusion properties of a molecule. Autocorrelation curves for molecules of different diffusional rates are illustrated in **Figure 9.12**. For a spherical molecule, the diffusion coefficient (D) is given by

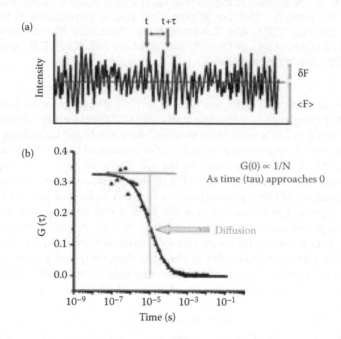

FIGURE 9.11 (a) Fluorescence correlation spectroscopy (FCS) data intensity stream indicating the average intensity, $<F(t)>$, the deviation in intensity from the average at a particular time point, $\delta F(t)$, and a time interval, t to $t + \tau$. (b) Autocorrelation curve indicating the characteristic diffusion time of the curve and the value of the autocorrelation function extrapolated to $\tau = 0$, i.e. $G(0)$, which is proportional to the reciprocal of the number of particles, N. (From D.M. Jameson et al., 2009. *Biophys. Rev.* 1: 105.)

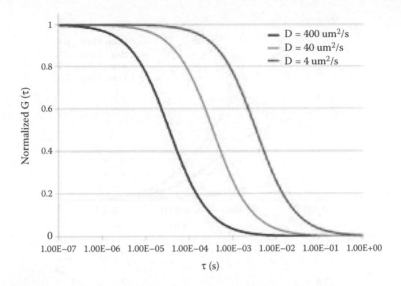

FIGURE 9.12 Simulation of the autocorrelation function of molecules with different diffusion coefficients with the G(0) normalized to 1. (From D.M. Jameson et al., 2009. *Biophys. Rev.* 1: 105.)

$$D = \frac{kT}{6\pi\eta r} \tag{9.2}$$

where k is Boltzman's constant $(1.38 \times 10^{23}$ J K$^{-1})$ r is the Stokes radius of the molecule, T is the temperature, and η is the solvent viscosity (this equation for translational diffusion can be compared to that for rotational diffusion— Equation 5.7). To give you a feeling for diffusional rates, in aqueous solution a small fluorophore, such as fluorescein, will have a diffusional coefficient near 400 μm^{2}s^{-1} (as mentioned above), whereas GFP, which has a molecular mass of ~27 kDa, will have a diffusional coefficient of ~90 μm^{2}s^{-1}.

Another very important feature of FCS data is that the method permits determination of the absolute concentration of the target fluorophore in the illumination/detection volume. Although this parameter may be obvious when dealing with homogeneous solutions of known concentration, it is extremely difficult to obtain when working with living cells and may be an important parameter in such studies. As the time interval, τ, goes to zero, the value of $G(\tau)$ goes to G_0, which approaches the reciprocal of the number of fluorophores, that is, $1/<N>$. This concept is illustrated in **Figure 9.13**, which shows actual data for different concentrations of rhodamine 110 in aqueous solutions.

Translational diffusion coefficients are related to the cube root of the molecular weight. Hence, the difference between the diffusional coefficients for a protein's monomer and dimer forms is only ~$2^{1/3}$ or ~1.26-fold, which is too small to distinguish accurately. Consequently, studies aimed at

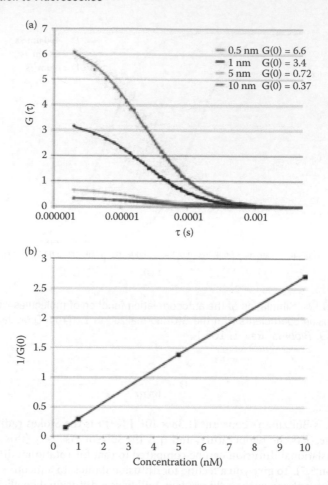

FIGURE 9.13 (a) Autocorrelation curves obtained for aqueous solutions of rhodamine 110 (Rh 110) at 22°C. The concentrations used are indicated on the figure. Squares indicate experimental data while solid lines represent the fit of the data to a Gaussian–Lorentzian point spread function (PSF) with the diffusion constant of 430 $\mu m2s^{-1}$. (b) Plot of the reciprocal of G(0) (which is proportional to <N>) as a function of Rh 110 concentration. Note that the calculated 1/G(0) values vary in proportion to the fluorophore concentration, as expected. (From D.M. Jameson et al., 2009. *Biophys. Rev.* 1: 105.)

determining a protein's oligomerization state cannot rely on classical FCS determinations of diffusional coefficients, unless the change in the oligomerization state is large. Two other FFS methods, however, can be applied to this problem. The first is based on a method developed independently, in 1999, by two groups from the United States and Germany, namely Chen et al. (1999, *Biophys. J.* 77: 553) and Kask et al. (1999, *Proc. Natl. Acad. Sci. USA* 96: 13756). These two groups named their methods the photon counting histogram (PCH) (Chen et al.) and the fluorescence-intensity distribution analysis (FIDA) (Kask et al.). In PCH and FIDA, the essential parameter is the inherent *molecular brightness* (B) of the fluorophore, that is, the actual counts per second per molecule (CPSM). This parameter will, of course, depend on both molecular factors (e.g., extinction coefficient or two-photon cross-section and quantum yield) and instrumental factors (optical path, laser power, detector efficiency, etc.). The idea is to measure the signal from a fluorophore standard, for example, monomeric GFP, and then, with the same instrumentation settings, switch to the sample. As shown in **Figure 9.14** (generously provided by Yan Chen and Joachim Mueller, University of Minnesota), the brightness parameter is proportional to the number of GFP molecules under observation.

The second method for studying protein oligomerization in cells is the number and brightness, or N&B method. Developed in Enrico Gratton's laboratory, the N&B approach can be considered the imaging equivalent of the PCH method; however, N&B does not require a nonlinear fit of the data. The average particle number, $<N>$, and particle brightness, B, are extracted directly from the image intensity data. The PCH method requires acquisition of a large number of photons at each point for reasonable precision of the oligomeric state of the target molecule and, as such, does not readily lend itself to image analysis. The N&B approach, however, although not as precise at each pixel in the image as the PCH method, allows for a rapid estimation of protein oligomerization size throughout an image. An example of the N&B approach is shown in **Figure 9.15**, in which the oligomerization states of the protein LRRK2 are shown in the plasma membrane of a cell (using TIRF microscopy).

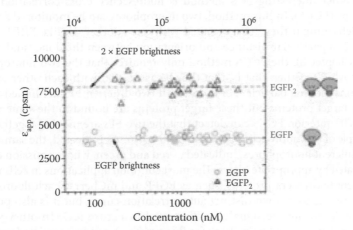

FIGURE 9.14 Molecular brightness of EGFP and EGFP2 as a function of average photon count rate and protein concentration. Note that the brightness levels are independent of concentration. Each data point represents the brightness measured in different cells expressing either EGFP or EGFP2. (From D.M. Jameson et al., 2009. *Biophys. Rev.* 1: 105; originally adapted from Y. Chen et al., 2003. *Proc. Natl. Acad. Sci. USA* 100: 15492.)

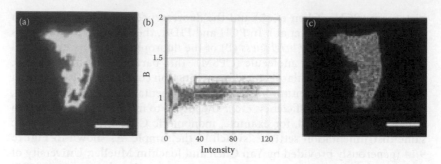

FIGURE 9.15 (a) TIRF intensity image of EGFP-LRRK2. (b) Brightness-versus-intensity plot showing the monomeric LRRK2 (green box), dimeric LRRK2 (red box), and the higher-order oligomers (blue box). (c) Brightness map highlighting the regions containing monomer (green), dimer (red), and higher-order oligomers (blue). Bars equal 17 mm. (From N.G. James et al., 2011. *Biophys. J.* 102: L43.)

The brightness versus intensity plot shown in **Figure 9.15b** allows one to isolate the pixels (from the original image shown in **Figure 9.15a**) in a particular brightness range and highlight those pixels in the image (shown in **Figure 9.15c**). The brightness (B) term is related to the true molecular brightness (ε) by: $B = 1 - \varepsilon$. The software for carrying out N&B analysis is available in the software package "Globals for Images" (aka "SimFCS") from the Laboratory for Fluorescence Dynamics. (http://www.lfd.uci.edu/).

Another interesting FCS method is fluorescence cross-correlation spectroscopy (FCCS). In this method, two fluorophores can be monitored and one can determine if they are associated with one another. Unlike FRET which requires a particular distance and orientation between the donor and acceptors (Chapter 8), the FCCS method only requires that the two fluorophores are moving together, that is, that they be associated with each other or with the same moving particle. For example, if two different FPs are linked to different target proteins but these target proteins are bound to the same vesicle in a cell's interior, FCCS can detect that the two FPs are moving together. An example of this approach is given in **Figure 9.16**. As indicated, the sample has two different fluorophores, indicated as red and green, whose emission can be separated by appropriate filters. The most common applications in cells utilize different fluorescent proteins, such as EGFP and mCherry. Each fluorophore will give rise to its own distinct autocorrelation curve, but it is also possible to cross-correlate the signals, as illustrated in **Figure 9.17**. In other words, the signal at one particular time for fluorophore 1 can be correlated with the signal at a different time for fluorophore 2. If the two fluorophores are in some way linked, then the resulting cross-correlated signal will show correlation, as indicated in the bottom panel of **Figure 9.17**.

Image correlation spectroscopy (ICS), was originally developed by Nils Petersen (Petersen 1986; Petersen et al. 1986) as an image analog of traditional FCS. In ICS, spatial autocorrelations are calculated from stacks of images

Sample

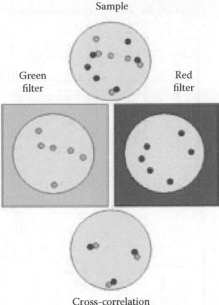

Green
filter

Red
filter

Cross-correlation

FIGURE 9.16 Illustration depicting a dual color cross-correlation scenario. A sample containing both "green" and "red" proteins can be viewed through green or red filters that pass only one color. In this case, since the number of proteins and their diffusion rate are shown to be the same, the autocorrelation curves for data acquired through either green or red filters should look similar (as shown in Figure 9.17). When the green and red signals are cross-correlated, however, only the dimers containing both green and red proteins will contribute to the cross-correlated signal (as shown in Figure 9.17). (From D.M. Jameson et al., 2009. *Biophys. Rev.* 1: 105.)

obtained in a time series. This method was extended in Enrico Gratton's lab to use a laser-scanning microscope to probe the time structure in images to spatially correlate pixels separated by microseconds (adjacent pixels in a line), milliseconds (pixels in successive lines), and seconds (pixels in different frames). This method is termed raster scanning image correlation spectroscopy (RICS). Many other variants of the ICS method have appeared, including TICS, ICCS, STICS, kICS, and PICS but I will not discuss these variants. Those readers wishing to venture into this acronym jungle can consult an excellent review by Kolen and Wiseman (2007, *Cell Biochem. Biophys.* 49: 141) or the aforementioned *Methods in Enzymology* volumes. The diffusion of a particle in a uniform medium can be described by the relation:

$$C(r,t) = \frac{1}{(4\pi Dt)^{3/2}} e^{-\left(r^2/(4Dt)\right)} \tag{9.3}$$

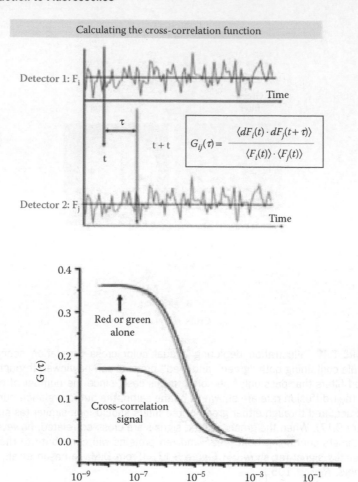

Calculating the cross-correlation function

Detector 1: F_i

Time

τ

$t + t$

$G_{ij}(\tau) = \dfrac{\langle dF_i(t) \cdot dF_j(t + \tau)\rangle}{\langle F_i(t)\rangle \cdot \langle F_j(t)\rangle}$

t

Detector 2: F_j

Time

Red or green
alone

Cross-correlation
signal

$G(\tau)$

Time (s)

FIGURE 9.17 (Top) Illustration of the method for calculating the cross-correlation function. (Bottom) Autocorrelation curves for the individual components and cross-correlated components shown in Figure 9.16.

where $C(r,t)$ represents the concentration of the particle at position r and time t, and D is the diffusion coefficient. In a RICS experiment, we are concerned with the spatial aspect. In this method, the spatial autocorrelation is similar to the time-dependent autocorrelation function carried out in traditional FCS, except that the correlation is carried out on different spatial points in the image, as illustrated in **Figure 9.18**. In this case, the autocorrelation is defined as

$$G(x, \Psi) = \frac{I(x,y)(I(x + x, y + \Psi)}{I(x,y)^2} \qquad (9.4)$$

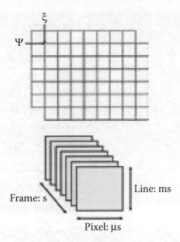

FIGURE 9.18 Sketch illustrating the multiple-shifting operation carried out to calculate a spatial correlation function. The time scale associated with each aspect of an image is also shown. Typically, 50 to 100 frames are required for a raster scanning image correlation spectroscopy (RICS) analysis. (From D.M. Jameson et al., 2009. *Biophys. Rev.* 1: 105.)

where ξ and ψ represent the spatial increments in the x and y directions, respectively, which are correlated. An example of a RICS analysis is shown in **Figure 9.19**, which gives examples of spatial correlation functions. As shown, an important aspect of the RICS method is that it permits the subtraction of immobile components and hence allows one to better quantify dynamic aspects of the system.

In fact, the theory and application of all of these FFS methods are somewhat complicated and, as always, readers wishing to learn more should refer to the primary literature. A series of lectures on these topics is also available on the website of the Laboratory for Fluorescence Dynamics (www.lfd.uci.edu) under the link for the Advanced LFD Workshop.

Fluorescence Recovery after Photobleaching (FRAP)

FRAP is one method which does not try to avoid photobleaching, but which actually embraces the destruction of fluorophores! In this method, the illumination spot, for example, from a laser source, is focused onto a specific spot on the sample. The method has been most commonly used with studies on cell membranes and such a scenario is illustrated in **Figure 9.20**. For a very brief time, the illumination intensity is increased markedly—such that the hapless fluorophores located in the illumination volume are promptly destroyed. Destroyed is actually a harsh assessment—let us just say that they are rendered nonfluorescent. The molecular mechanisms which result in photobleaching of an excited fluorophore are diverse, but perhaps the best characterized bleaching mechanism involves the reaction of the excited molecule with oxygen, which

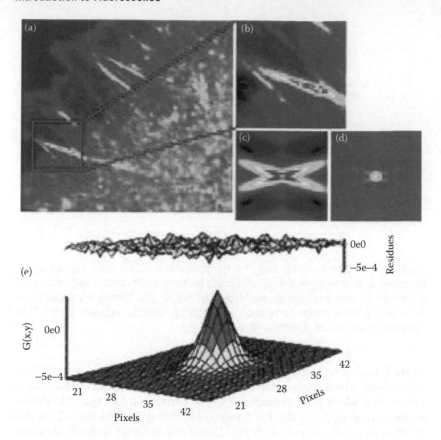

FIGURE 9.19 (a) Image of a CHO-K1 cell expressing paxillin–EGFP. (b) 64 × 64 subframe in the cytosolic part of a focal adhesion structure. Spatial autocorrelation before (c) and after (d) the subtraction of immobile structures. (e) Fit of the spatial correlation function in d. The diffusion coefficient in this cell region is 0.49 $\mu m^2 s^{-1}$. (From D.M. Jameson et al., 2009. *Biophys. Rev.* 1: 105; originally adapted from M.A. Digman et al., 2005. *Biophys. J.* 89: 1327.)

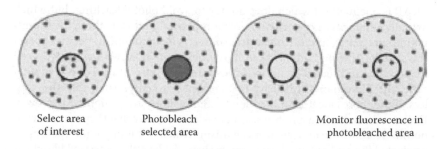

Select area of interest   Photobleach selected area   Monitor fluorescence in photobleached area

FIGURE 9.20 Illustration of the FRAP method.

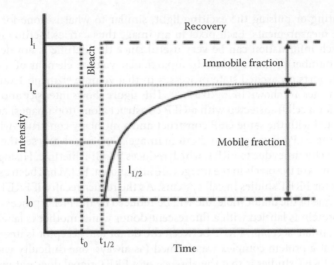

FIGURE 9.21 Illustration of a FRAP bleaching and recovery process.

may result in loss of conjugation (see Chapter 2) and hence, loss of absorptivity at the illumination wavelength. Regardless of the precise photobleaching mechanism, though, the result is that there are suddenly fewer fluorophores in the illuminated volume and hence the recorded signal can be dramatically reduced. Assuming that the photobleached molecules are rendered permanently *hors de combat*, the only hope for a recovery of the fluorescent signal in the illuminated volume is for more fluorophores to enter the area which was bleached, as depicted in **Figure 9.20**. The recovery of the fluorescence can be monitored and the rate and extent of recovery (**Figure 9.21**) can provide information on the diffusional processes involved as well as what fraction of the target fluorophores are mobile. Fluorescence loss in photobleaching (FLIP) is a technique related to FRAP in which multiple photobleaching steps are used to largely deplete a large portion of a cell (e.g., nuclear regions or cytoplasmic regions) of target fluorophores (typically GFP-linked proteins). I should add that photobleaching is not the only issue that concerns those working with fluorescence microscopy. Phototoxicity may also be a problem since photogenerated reactive oxygen species can cause significant damage to cells, as can fluorescent-independent processes such as light-induced cell heating.

Fluorescence Lifetime Imaging Microscopy (FLIM)

Interestingly, fluorescence lifetimes were measured through a microscope as early as 1959 by B.D. Venetta, who measured the lifetime of proflavin in the nuclei of tumor cells using a home-built frequency domain instrument. This early instrument, however, did not provide lifetime imaging. Modern FLIM instruments, which provide lifetime measurements for each pixel in the image, have found increasing applications in cell studies. FLIM can be implemented using either frequency or time-domain approaches, which involves either modulating or pulsing the exciting light, similar to what is done for *in vitro* lifetime measurements. Each pixel in an image thus carries lifetime information. Such information can be very useful since the lifetime is not dependent on the number of fluorophores in a particular volume element of a cell, but rather reports on excited state processes in that volume element. Examples of FLIM images are shown in **Figure 9.22**. This figure shows intensity and lifetime images for a cell transfected with a GFP construct alone (top images) and a cell transfected with the same GFP construct and a mCherry construct, which can bind to the GFP-tagged protein (bottom image). One can clearly see the areas of decreased lifetimes due to FRET, which reduces the GFP lifetime. Histograms of the lifetimes of the pixels in the images are also shown. FLIM has been especially popular for FRET studies in cell systems. As the reader recalls, if FRET occurs, the lifetime of the donor molecule will be reduced by the FRET process. Hence, if one protein is labeled with a fluorescent donor while another is labeled with an appropriate acceptor, FRET between the donor and acceptor is unequivocal proof that a protein complex was formed (as always, one difficulty with these types of FRET studies is that the absence of a FRET signal does not prove that the protein complex did not form, since the donor and acceptor dipoles may be too far apart or they may not be oriented properly for FRET to occur). Similarly, in the case of FRET-based biosensors (discussed in more detail in Chapter 10), binding of the properly designed biosensor to the target results in the change of disposition of the donor and acceptor parts of the biosensor and a change in the donor lifetime.

In the case of FLIM on live cell systems, the situation is complicated by the fact that any particular pixel composing the image will not typically have

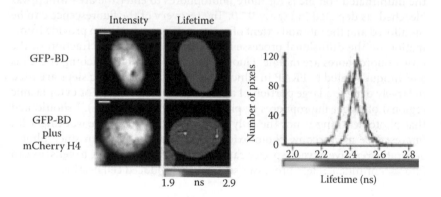

FIGURE 9.22 Intensity and FLIM images of GFP-BD (BD refers to the double bromodomain of a histone acetyltransferase) expressed alone as a control (upper panel) or with mCherry-H4 (H4 refers to an acetylated peptide from histone H4) as a cotransfection (lower panel) in HEK293 live cells. The corresponding lifetime histograms are presented in the right panel (black for control and red for cotransfection). White arrows show two chromatin domains in which GFP-BD mean lifetime decreases significantly. Scale bars are 2 microns. (Modified from S. Padilla-Parra et al., 2008. *Biophys. J.* 95: 2976.)

more than 500–1000 photons contributing to the signal. Hence, extracting a precise lifetime is problematic, especially if multiple lifetime components are present. Also, one may have to contend with autofluorescence, which makes a rigorous analysis of a donor lifetime very difficult to achieve. For this reason, the phasor approach (discussed in Chapter 6) to FLIM has found increasing use. Using the FLIM analysis software in the Globals for Images software package (aka Sim FCS) developed by Enrico Gratton and available from the Laboratory for Fluorescence Dynamics (www.lfd.uci.edu), one can construct a phasor diagram from the FLIM image. Then, one can place a cursor on any point in the phasor diagram and the pixels in the image corresponding to those phasor values will be illuminated. This principle is illustrated in **Figure 9.23**. In this example, microchannels in the shape of the letters L, F, and D were filled with either fluorescein (L), rhodamine B1 (D), or a mixture of these two dyes (F). The phasor diagram corresponding to the images are shown in the bottom row of the figure. Placing the cursor over the phasor region on the left side of the plot illuminates the pixels (pink) in the image corresponding to those phasor points (the letter L), in this case due to fluorescein. Correspondingly, placing the cursor over the phasor points due to rhodamine illuminates the pixels in the letter D. Finally, placing the cursor over the points corresponding to the mixture illuminates the pixels in the letter F.

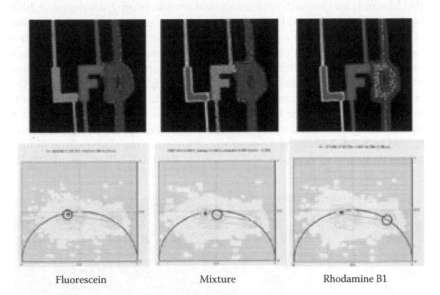

Fluorescein   Mixture   Rhodamine B1

FIGURE 9.23 Illustration of the phasor plot program in the simFCS package (www.lfd.uci.edu). The three letters (L,F,D) are filled with fluorescein (L), rhodamine (D) and a mixture of the two dyes (F). The phasor plots corresponding to the overall image are shown beneath each letter. As the cursor is placed over the different phasor points different letters light up demonstrating which pixels contained the chosen lifetime data. (From D.M. Jameson et al., 2013. *Methods* 59: 278. The author would like to thank Enrico Gratton for the original figure.)

Super-Resolution Techniques

In the last decade or so, several approaches, both instrumental and biological, have been developed to greatly improve spatial resolution in microscopy. As the reader may know, the limit to the angular resolution of a microscope was traditionally considered to be the so-called diffraction limit or Abbe limit (in honor of Ernst Abbe), approximately half of the wavelength of the light utilized. In other words, if one illuminates a sample with 500 nm light, then the resolution attainable is approximately 250 nm. In 1991, Stefan Hell developed a practical method of what is termed 4Pi microscopy (which shall not be discussed), which eventually led to saturation emission depletion (STED) microscopy. In the STED approach, a wavelength of light, displaced from the excitation wavelength, is used to stimulate emission from fluorophores in the region, typically donut-shaped, surrounding a central area (the donut hole). This technique results in a dramatic improvement in spatial resolution compared to conventional confocal microscopy, as shown in **Figure 9.24**. STED methods are still evolving, as specialized dyes and excitation sources continue to be developed. Newer implementations of the STED approach include ground state depletion (GSD), saturated structured illumination microscopy (SSIM), and reversible optically linear fluorescence transitions (RESOLFT).

Biological approaches to super resolution began with the PALM approach. PALM stands for photoactivatable localization microscopy, developed in 2006 by Betzig et al. (*Science*, 313: 1642). The essential idea is to limit the number of emitting molecules in a region of interest. To accomplish this goal, a photoactivatable fluorescent protein (PA-FP), based on GFP, was designed such that when it was biosynthesized, it was not fluorescent until light of the appropriate wavelength was introduced. This activating light then converted the target protein into a normal fluorescent state. If the activating light is of sufficiently low intensity, the PA-FP molecules can be turned on over time such that the image will build up slowly. As a fluorescent spot appears, the center of that spot can be accurately determined knowing the shape of the illumination point spread function, and hence, the spatial details of the molecular

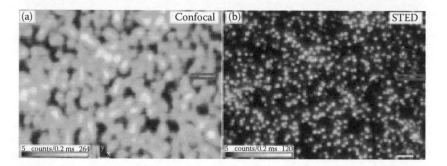

FIGURE 9.24 Comparison of a conventional confocal microscope image (a) and a STED image (b) of dispersed 40 nm beads. Scale bar is 500 nm. (From K.I. Willig et al., 2006. *New J. Phys.* 8: 106.)

structure containing the PA-FPs can be revealed. The approach known as stochastic optical reconstruction microscopy or STORM, is similar to PALM in that it also utilizes sequential activation and temporal resolution of fluorescent molecules, although in its original implementation by Rust et al. (*Nature Methods* (2006) 3: 793) it used cyanine dyes, not fluorescence proteins.

Single-Molecule Fluorescence

Since entire books have been written on single-molecule fluorescence methods, I will not cover them in any significant detail. I mainly want to point out the fact that although these approaches look at single molecules, there are still plenty of photons involved! The basic hurdle to overcome in single molecule fluorescence is first to be able to localize the fluorophore such that it will not diffuse out of the field of view, and second, to be able to collect as many photons as possible from that fluorophore. Typically, to prevent diffusion during the measurement, the fluorophore system is tethered to a surface (e.g., using biotin–streptavidin coupling) or embedded in a matrix. To collect as many photons as possible, it is necessary to use fluorophores that are intrinsically very bright and which are not prone to photobleaching. Given the fact that fluorescence lifetimes are typically in the range of nanoseconds, one can see that it is possible to excite an individual fluorophore a great many times, even in periods as short as a millisecond. The real trick, of course, is using a fluorophore that will not photobleach readily; some of these are discussed in Chapter 10. A classic use of single molecule fluorescence is to observe FRET from an individual donor molecule. For example, **Figure 9.25** illustrates

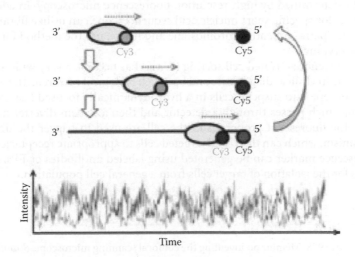

FIGURE 9.25 A single helicase molecule moving on a strand of DNA, directionally powered by ATP hydrolysis. The helicase is labeled with the donor (Cy3), while the DNA is labeled with the acceptor (Cy5), and the intensities from the individual fluorophores can be recorded, as shown in the intensity trace (bottom). (The author would like to thank Taekjip Ha for this figure.)

single molecule FRET measurements carried out on labeled DNA molecules, specifically, a study of a helicase molecule moving on a strand of DNA, directionally powered by ATP hydrolysis. The helicase is labeled with the donor (Cy3), while the DNA is labeled with the acceptor (Cy5), and the intensities from the individual fluorophores can be recorded, as shown in the bottom of **Figure 9.25**. One notes that the signals from the donor and acceptor are anticorrelated, that is, when the donor fluorescence is high, the acceptor's is low, and vice versa. Upon reaching the end of the DNA (as evidenced by the increase in FRET), the helicase reappears at the beginning of the DNA strand to repeat the process.

Miscellaneous

There are, of course, other techniques, important in biology and medicine, which utilize fluorescence. However, given the introductory nature of this book and space limitations, I shall not discuss them. I will only point out two of the fluorescence-based methods in widespread use in biomedicine.

Fluorescence *in situ* hybridization (FISH) is a method used to localize and quantify specific nucleic acid sequences. Typically, fluorescence probes are incorporated into short sequences of single- or double-stranded DNA or RNA, either by direct labeling of individual nucleotides followed by polymerization or by direct labeling of the nucleic acid polymers. These probes can then be introduced into a cell where they will bind to complementary nucleic acid sequences (in the case of double-stranded probes, a denaturation and hybridization step may be included). The location of these sequences in the cell can then be determined by high-resolution fluorescence microscopy. In addition to probes for specific short nucleic acid sequences, one can utilize libraries of probes to "paint" an entire chromosome and hence, use the method for spectral karotyping.

Fluorescence activated cell sorting (FACS) has become a very wide-spread technique in clinical diagnostics and basic cell biological research. The underlying concept is to suspend cells in a hydrodynamically focused fast moving stream, which passes through a detector and then a stream diverter mechanism. The fluorescent properties of the cell are used to trigger the diverter mechanism, which can then send targeted cells to appropriate receptacles. The fluorescence marker can be generated using labeled antibodies or FPs. FACS allows for the isolation of target cells from a general cell population.

Additional Reading

M. Minsky, 1988. Memoir on inventing the confocal scanning microscope. *Scanning* 10: 128–138.

W. Denk, J.H. Strickler, and W.W. Webb, 1990. Two-photon laser scanning fluorescence microscopy. *Science* 248: 73–76.

J.B. Pawley (ed.), 2006. *Handbook of Biological Confocal Microscopy*, (3rd edn). Springer, New York, NY.

D.L. Kolin and P.W. Wiseman, 2007. Advances in image correlation spectroscopy: Measuring number densities, aggregation states, and dynamics of fluorescently labeled macromolecules in cells. *Cell Biochem Biophys* 49: 141–164.

V.R. Caiolfa, M. Zamai, G. Malengo, A. Andolfo, C.D. Madsen, J. Sutin, M.A. Digman, E. Gratton, F. Blasi, and N. Sidenius, 2007. Monomer dimer dynamics and distribution of GPI-Anchoredu PARAre determined by cell surface protein assemblies. *J. Cell Biol.* 179: 1067–1082.

P.R. Selvin and T. Ha (eds.), 2008. *Single-Molecule Techniques: A Laboratory Manual*, Cold Spring Harbor Laboratory Press, Cold Spring Harbor, NY.

D.M. Jameson, J.A. Ross, and J.P. Albanesi, 2009. Fluorescence fluctuation spectroscopy: Ushering in a new age of enlightenment in cellular dynamics. *Biophys. Rev.* 1: 105–118.

S.Y. Tetin (ed.), 2013. *Methods in Enzymology*, Vols. 518 and 519. Elsevier, Inc., Amsterdam, the Netherlands.

10

Fluorophores

IT IS NOT SURPRISING THAT this chapter on fluorophores is the longest in the book. Without fluorophores, I would be out of business! Before diving in though, I want to ask the reader not to be disappointed if her or his favorite fluorophores are not discussed. As you can imagine, the list of fluorescent molecules, even just those commercially available, is far too long to allow a comprehensive treatment. My goal, rather, is to cover many of the better known categories of fluorophores and to illustrate some of their applications.

Where Do Fluorophores Come From?

Some fluorophores occur naturally in the plant or animal world, such as quinine, which has played an important role in the development of the fluorescence field (Chapter 1). Many such natural fluorophores, or "intrinsic fluorophores," including the aromatic amino acids, are very useful in modern research and will be discussed here. Yet, most modern fluorescence studies would not be possible without fluorescent molecules designed and synthesized in a laboratory. Indeed, the rapid development of the biological fluorescence field owes much to synthetic organic chemistry, as well as to modern molecular biology and the commercialization of instrumentation. Most beginning fluorescence practitioners do not have to synthesize their own fluorophores, since literally thousands are now available. Yet, even as late as the 1970s—when I was a graduate student with Gregorio Weber—only a handful of fluorescence probes were commercially available. Consequently, for some of my studies, I had to synthesize the probe, build the instrument, and walk 5 miles uphill each day through the snow to the lab (OK—maybe I didn't actually have to walk 5 miles uphill through the snow—but the rest is true)! Much of the usefulness of fluorescence observations for the chemist and biologist stems from the fact that most of the molecules that absorb light in the ultraviolet and visible regions of the spectrum do not emit conspicuous fluorescence. In fact, the difference between fluorescent and "nonfluorescent" molecules is often quantitative rather than qualitative, and may be more precisely characterized by the fluorescence quantum yield, as discussed in Chapter 6. For example, nucleosides and nucleotides actually fluoresce upon deep UV excitation (~260 nm), but their quantum yields are extremely low, on the order of 10^{-4}, which, of course, severely limits their practical use as fluorophores. I should mention that, unlike some authors, I may use the terms fluorophores, probes, dyes, labels, or indicators interchangeably since these terms imply particular uses for the fluorophores being discussed. To paraphrase Shakespeare, "A fluorophore by any other name will glow as brightly!"

How Does One Choose a Fluorophore?

Occasionally, I am asked, "What fluorophore should I use to study my system?" To address this question one must, of course, first know exactly what the system is and what type of information is being sought. For example, if one is studying a protein *in vitro* and wishes to quantify a dimer to monomer transition, one may want to attach a fluorophore covalently to the protein, either using an amine or perhaps a sulfhydryl reactive probe. Then, one could dilute the protein over a suitable range and use polarization/anisotropy to see if dimer dissociation takes place, and the effects of ligands on the equilibrium. In this case, the lifetime of the fluorophores will be important, as we learned from Chapter 5 on polarization/anisotropy. If one wishes to follow protein unfolding or refolding, the intrinsic protein fluorescence (more on this later) could be used. If one's purpose is to quantify a particular ion, for example, calcium, either *in vitro* or in living cells, then appropriate probes are available to accommodate the precise ion and binding constant involved. Yet another common approach is to use fluorescence to probe the physical state of a lipid membrane, in which case a lipid-soluble probe with appropriate fluorescence properties is required. Fluorescent molecules have been designed to probe a wide range of physical or chemical conditions as depicted in **Figure 10.1** (adapted from Bernard Valuer's book *Molecular Fluorescence*). In this chapter, we shall discuss some of these diverse probes. Although the chemical structures for numerous fluorophores shall be presented, absorption and emission spectra are presented for only a few of these. There is just not enough room in this introductory book to give a comprehensive guide to all of the spectral properties of every fluorophore mentioned. Such additional information is readily available on the Internet, for example, on the websites of the companies selling fluorophores, such as www.invitrogen.com, www.promega.com, www.piercenet.com, www.setabiomedicals.com, and www.activemotif.com, among others. A database on fluorescence probes is also available at www.fluorophores.tugraz.at.

Intrinsic Probes

By the term "intrinsic," we mean fluorophores that occur naturally in the particular system we wish to study. In biological systems, the most common

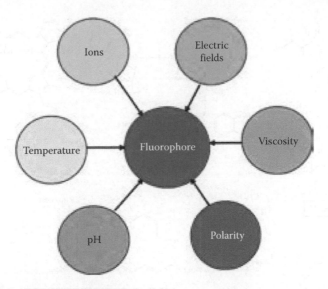

FIGURE 10.1 Diagram of various physical and chemical conditions to which fluorescence probes can respond.

endogenously fluorescent molecules include the pyridinic and flavin coenzymes (NADH, FAD, FMN), some porphyrins including chlorophyll (not iron-bound porphyrins, however, since iron will quench the fluorescence), pyridoxines, pteridines, and proteins. A few other intrinsic fluorophores are the Wye-base (wybutosine) in some transfer RNAs and some endogenously fluorescent lipids such as parinaric acid. Collagen and elastin, two proteins present in connective tissues, also exhibit fluorescence due to their cross-linked products. Proteins are probably the most widely studied naturally fluorescent system. The only amino acids that are naturally fluorescent are tryptophan, tyrosine, and phenylalanine. The study of intrinsic protein fluorescence is so widespread that it deserves its own treatment and shall be covered in Chapter 11. The structures of several endogenously fluorescent molecules are shown in **Figure 10.2**.

Extrinsic Probes

By the term "extrinsic," we generally mean any fluorophore which does not occur in nature. It is interesting that the majority of man-made fluorophores are actually based on only a handful of chemical core groups. These core groups include xanthene, naphthalene, cyanine, BODIPY, coumarin, NBD, oxazine, and bimane; examples of probes from most of these families will be shown later in this chapter. It is manifestly impossible for me to provide a comprehensive list of available extrinsic fluorescence probes, that is, probes available commercially, let alone probes synthesized and utilized but not yet commercially available. Even a casual perusal of the literature will show that there are many hundreds, perhaps thousands, of known fluorophores. The

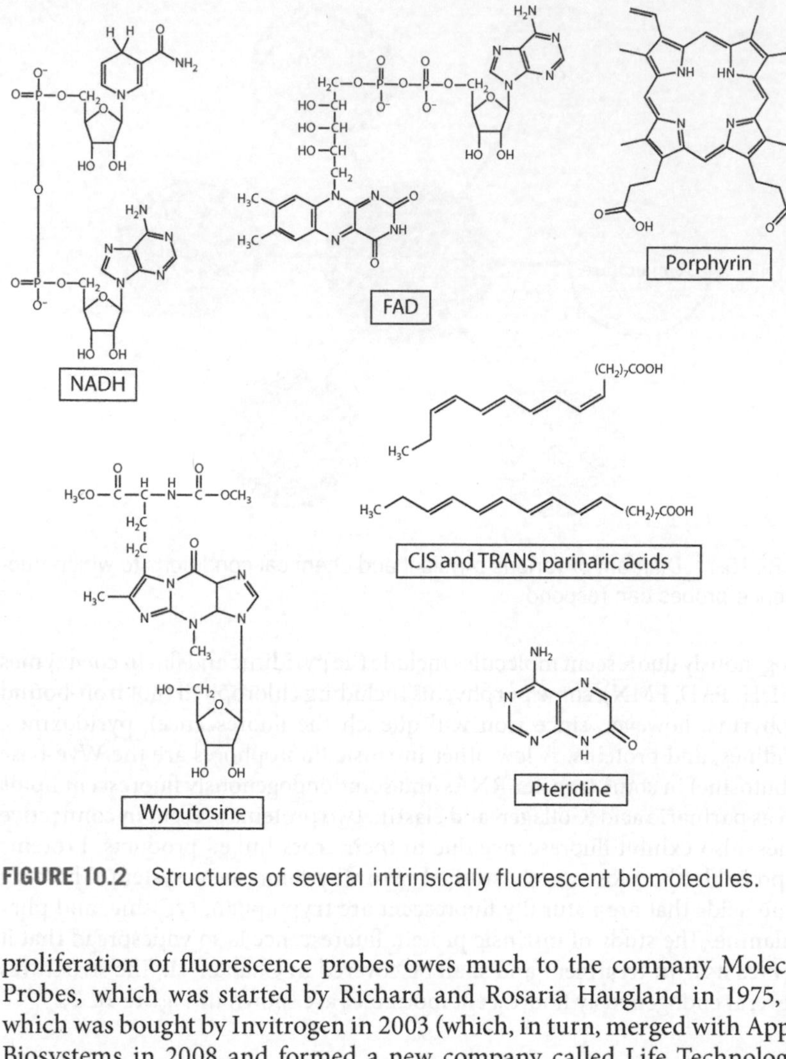

FIGURE 10.2 Structures of several intrinsically fluorescent biomolecules.

proliferation of fluorescence probes owes much to the company Molecular Probes, which was started by Richard and Rosaria Haugland in 1975, and which was bought by Invitrogen in 2003 (which, in turn, merged with Applied Biosystems in 2008 and formed a new company called Life Technologies). In recent years, other "designer fluorophore" companies have appeared and the commercial availability of so many probes has greatly contributed to the widespread popularity of fluorescence. The field has come a long way indeed from the days when beginning graduate students had to synthesize their own probes (although thankfully there are still some intrepid chemists around)!

The first extrinsic probe (at least the first one to achieve widespread use) was fluorescein, made in 1871 by Adolf von Bayer (as mentioned in Chapter 1). The first extrinsic fluorescein label was made by Albert Coons and his colleagues in 1941, who labeled antibodies with fluorescein isocyanate, thus giving birth to the field of immunofluorescence. J. L. Riggs and colleagues first reported the synthesis of fluorescein isothiocyanate (FITC; **Figure 10.3**) in 1958, which they synthesized to circumvent problems inherent in the

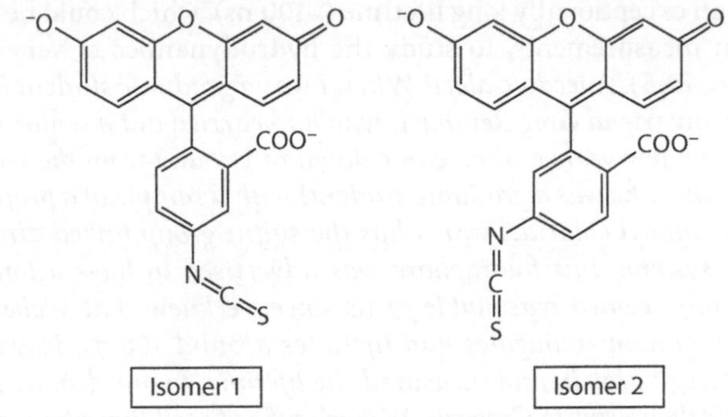

FIGURE 10.3 Structures of fluorescein isothiocyanate (FITC) isomer 1 and isomer 2.

isocyanate derivative, including the difficulty of its synthesis and its instability. FITC became, arguably, the most popular fluorescent label of all time. I note that FITC is available in two common isomers; isomer 1 or 2, which have the isothiocyanate group on carbon 4 or 5 of the benzene ring, respectively. Isomer 1 is easier to purify and hence, is usually less expensive.

By 1951, Gregorio Weber had begun to develop methods which would allow him to study proteins not containing intrinsic fluorophores such as FAD or NADH (the fluorescence of the aromatic amino acids had not yet been discovered). To this end, he invested considerable time and effort in synthesizing a fluorescent probe, which could be covalently attached to proteins and which possessed absorption and emission characteristics appropriate for the instrumentation available in post-war England. The result of two years of effort was the still popular probe 1,5-dimethylaminonaphthalene sulfonyl chloride or dansyl chloride (**Figure 10.4**). Using this probe, Weber initiated the field of quantitative biological fluorescence.

During the decades after the introduction of dansyl choride, relatively few new probes were synthesized (with some notable exceptions that will be discussed). In 1967, Weber covalently labeled proteins with pyrenebutyic acid, a

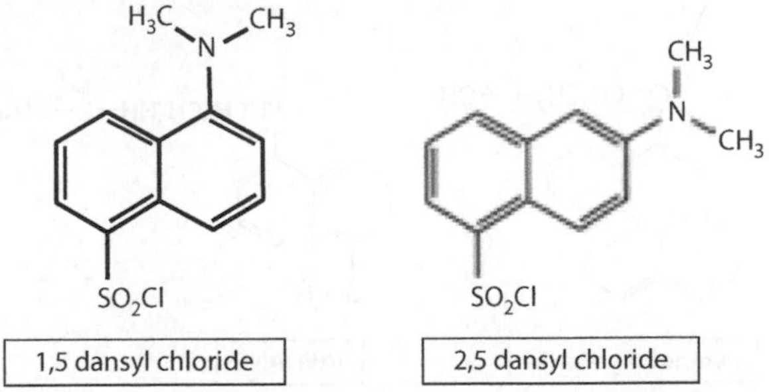

FIGURE 10.4 Structures of 1,5- (and 2,5-) dimethylaminonaphthalene sulfonyl chloride (dansyl chloride).

195

probe with an exceptionally long lifetime (>100 ns), which could be used, with polarization measurements, to study the hydrodynamics of very large proteins (**Figure 10.5**). *Anecdote alert! When I was a graduate student in Gregorio Weber's lab, my friend Greg Reinhart, who had carried out a senior thesis project in our lab a few years earlier, came down to Urbana from the University of Wisconsin (where he was a graduate student) with a sample of a protein labeled with pyrenesulfonyl chloride, which has the sulfur group linked directly to the pyrene ring system. This fluorophore was advertised to have a long lifetime, which certainly seemed reasonable to us since we knew that Weber's pyrenebutyric acid–protein conjugates had lifetimes around 100 ns. However, when we labeled Greg's protein and measured the lifetime, to our dismay it was only 4 or 5 ns. We then went to Gregorio Weber's office to tell him about our results, but before we had a chance to report the lifetime, Weber looked at the structure of the probe and said that one would expect a lifetime of 4 or 5 ns. Needless to say, Greg and I were amazed and asked Weber how he knew the lifetime. He replied that the reason he used pyrenebutyric acid was to have a chain of methylene groups between the ring system and the electron-withdrawing carbonyl group, since if an electron-withdrawing group was linked to the aromatic system (as was the case with our pyrenesulfonyl moiety), it would destroy the symmetry, which leads to the forbidden nature of the electronic transition and the long lifetime! While writing this anecdote, I was struck by the fact that in those days, we did not have Google available to quickly check for information—instead we had Gregorio Weber! The advantage of Weber over Google was that he always provided the correct information.* In 1973, Weber introduced N-(iodoacetylaminoethyl)-5-naphthylamine-1-sulfonic acid (1,5-IAEDANS), the first probe designed to react with sulfhydryl residues, specifically cysteines in proteins (**Figure 10.5**). As long as we are discussing probes synthesized by Gregorio Weber, we must mention the family of 2,6 naphthalene probes, which he designed to have a large excited state dipole moment, which would confer a large environmental sensitivity (since relaxation of surrounding dipoles would result in a shift of the probe's excited state to lower energies and thus red-shifted emission; more on this topic later). The probes, Prodan, Laurdan, and Danca are shown in **Figure 10.6** and these will be discussed in more detail later in this chapter. An increasingly popular fluorophore, which forms

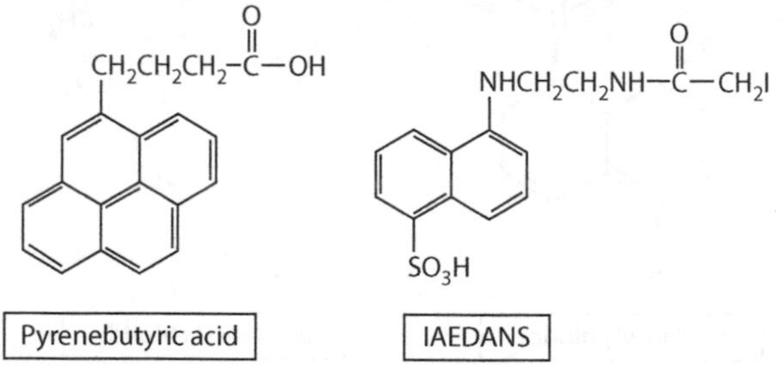

Pyrenebutyric acid

IAEDANS

FIGURE 10.5 Structures of pyrenebutyric acid and N-(iodoacetylaminoethyl)-5-naphthylamine-1-sulfonic acid (IAEDANS).

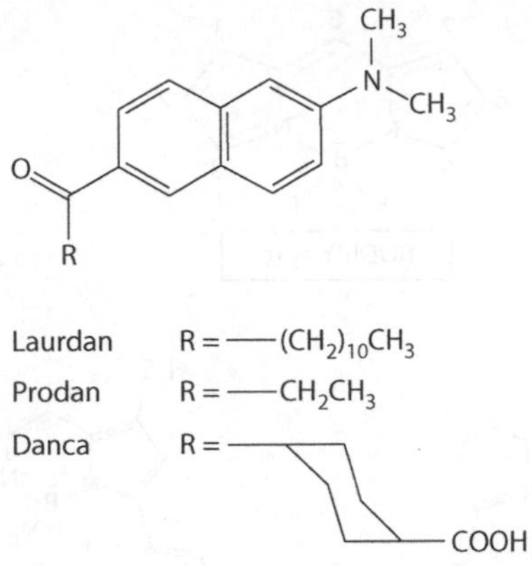

FIGURE 10.6 Structures of the family of 2,6-naphthalene probes.

the core of myriad probes, is 4,4-difluoro-4-bora-3a,4a-diaza-*s*-indacene, or BODIPY. Originally described by Alfred Treibs and Franz-Heinrich Kreuzer in 1968 (Justus Liebigs Annalen der Chemie, 1968, 718: 208), BODIPYs have useful spectroscopic properties, such as large extinction coefficients and quantum yields and environmental sensitivities, that recommend their use in many systems, including biosensors. The BODIPY "core," along with several BODIPY-based fluorophores, is shown in **Figure 10.7**. Commercially available BODIPY compounds are generally named to reflect their emission properties, specifically, by either recalling their spectral similarities to well-known fluorophores such as fluorescein (Fl), rhodamine-6-G (R6G) or tetramethylrhodamine (TMR) or by explicitly giving the excitation and emission maxima, that is, BODIPY 650/665. 7-nitrobenz-2-oxa-1,3-diazol-4-yl (NBD) (excitation maximum ~450 nm, emission maximum ~530 nm) is another popular probe often used to modify biomolecules such as carbohydrates and lipids (**Figure 10.8**). NBD has also been functionalized to make it reactive with proteins, for example, as succinimidyl ester or maleimide forms (these linkage chemistries will be discussed later in this chapter).

Photostability

In 1999, Richard Haugland and colleagues designed a series of probes, named the Alexa series (after Haugland's son, Alex), which exhibited greatly improved photostability compared to many commonly used probes, such as fluorescein. One of the main motivations for development of photostable probes was the increasing popularity of fluorescence microscopy. Probes such as fluorescein, which were fine for most *in vitro* studies, were rapidly photobleached in the intensely bright focal spots of microscope objectives. The new Alexa probes were based on sulfonation of commonly used fluorophores such as xanthenes, coumarins, and cyanines. Alexa probes (available from Invitrogen) are available with a wide range of excitation and emission wavelengths and have

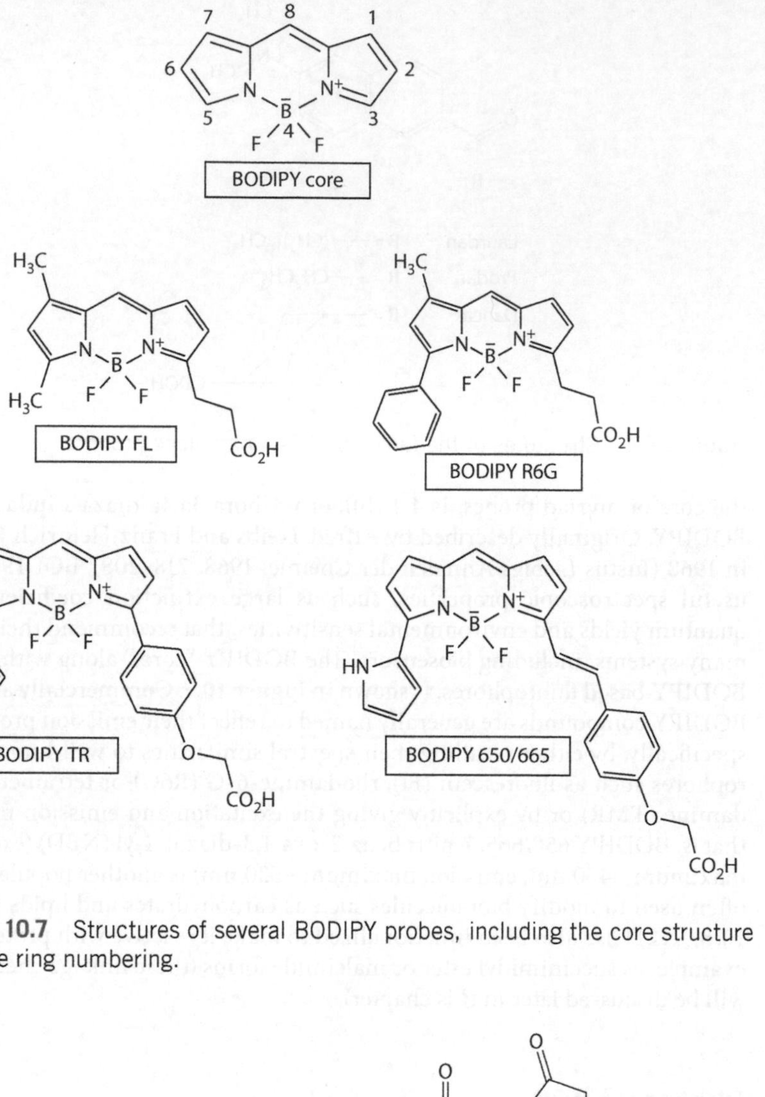

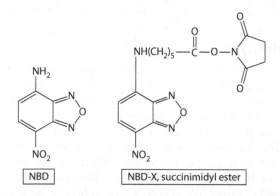

FIGURE 10.7 Structures of several BODIPY probes, including the core structure with the ring numbering.

FIGURE 10.8 Structure of NBD and NBD-X, a succinimidyl ester of NBD.

become a popular choice for those interested in covalently labeling proteins. They are available with either amine or sulfhydryl reactive chemistries. Some members of the Alexa series are shown in **Figure 10.9**. Numerous other probes with enhanced photostability have been developed in the last decade, such as the DyLight Fluor series (available from Thermo Fisher Scientific) the ATTO series (available from ATTO-TEC Gm) and the FluoProbes series (available

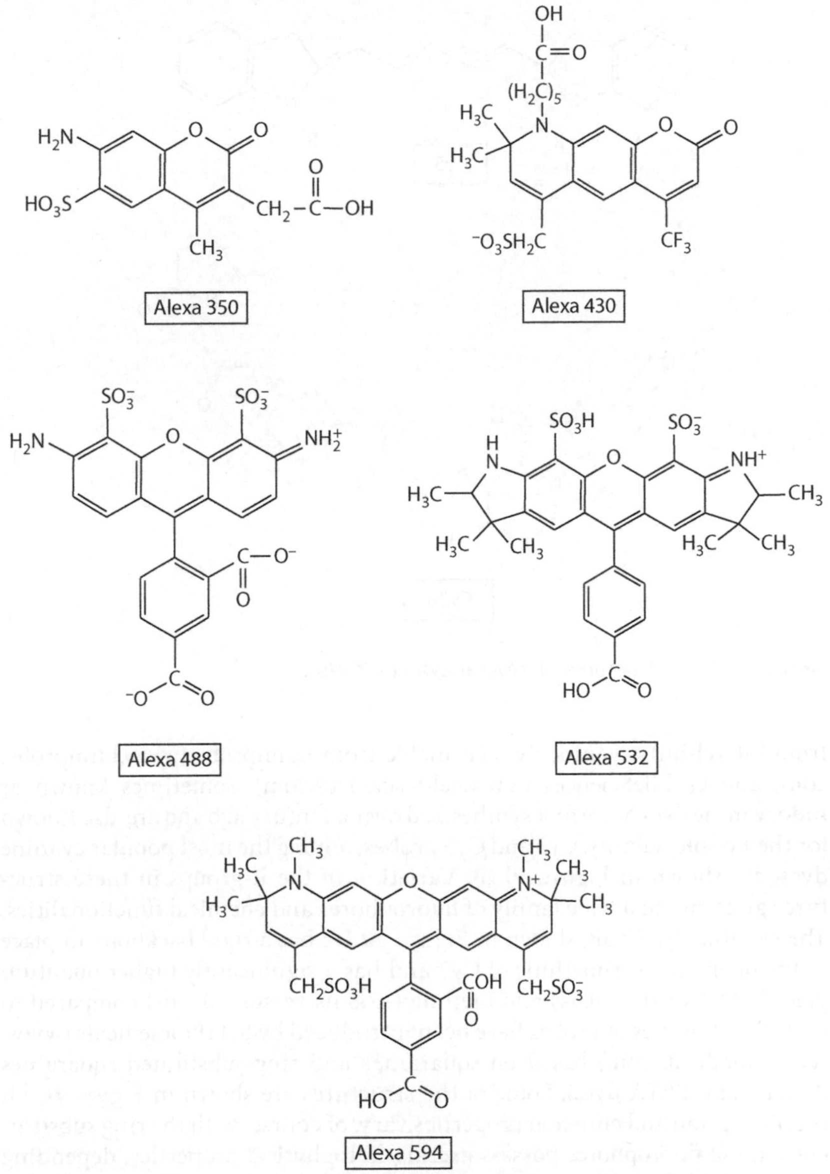

FIGURE 10.9 Structures of several Alexa fluorophores.

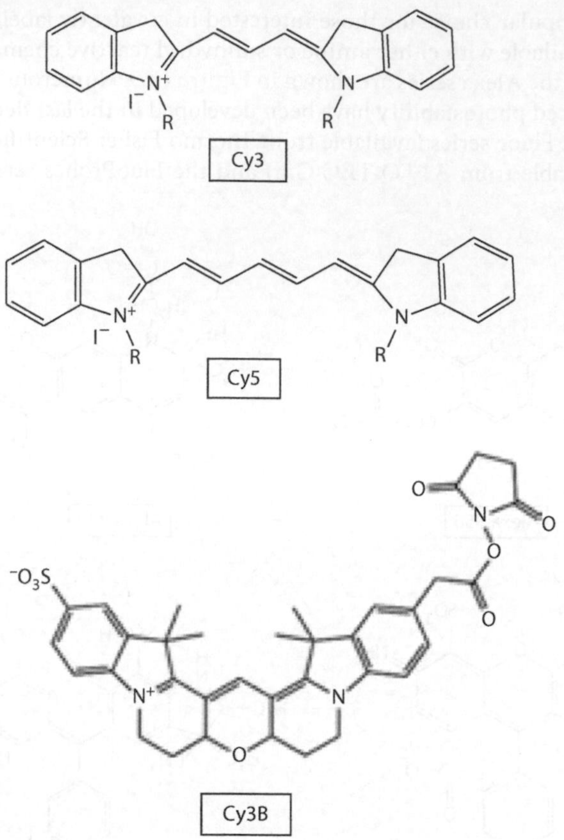

FIGURE 10.10 Structures of several cyanine dyes.

from Interchim). Cyanine dyes (available from Lumiprobe (www.lumiprobe. com) and GELifeSciences (www.gelifesciences.com), sometimes known as indocyanine dyes, were first synthesized over a century ago and are also known for their photostability. Cy3 and Cy5 probes, among the most popular cyanine dyes, are shown in **Figure 10.10.** Variation of the R-groups in these structures gives rise to a wide family of fluorophores and chemical functionalities. The cyanine dye Cy3B, shown in **Figure 10.10,** has a rigid backbone in place of the open chain trimethine of Cy3 and has a significantly higher quantum yield (~0.67 versus ~0.04) and lifetime (~2.8 ns versus ~0.3 ns) compared to Cy3. New families of probes have been introduced by SetaBiomedicals (www. setabiomedicals.com) based on squaraines and ring-substituted squaraines (Square and SETA dyes). Some of the structures are shown in **Figure 10.11;** the absorption and emission properties vary, of course, with the ring substituents. These fluorophores possess useful photophysical properties, depending on the exact derivatives on the ring systems, such as very large extinction coefficients, photostability, long-lifetimes (for some probes) and near-infrared

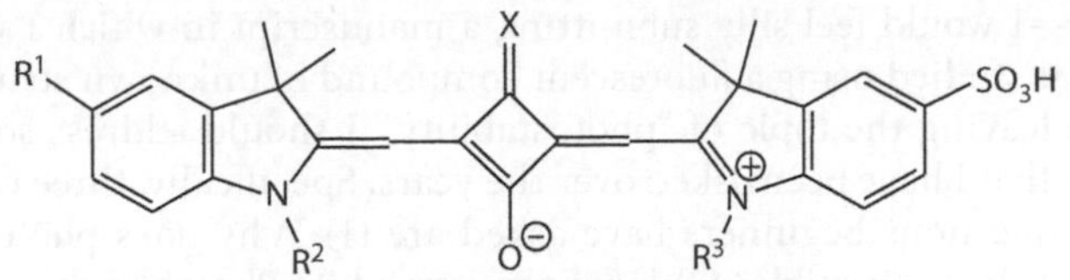

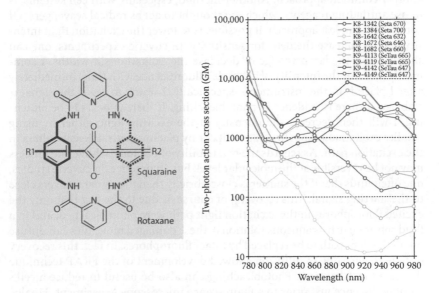

$$X = S, R^1 = H, R^2 = Me, R^3 = (CH_2)_5COOH$$
$$X = C(CN)_2, R^1 = H, R^2 = Me, R^3 = (CH_2)_5COOH$$
$$X = O, R^1 = H, R^2 = Me, R^3 = (CH_2)_5COOH$$
$$X = S, R^1 = SO_3H, R^2 = R^3 = (CH_2)_5COOH$$

FIGURE 10.11 Structure of several squaraine-based fluorophores. (The author would like to thank Ewald Terpetschnig from Seta Biomedicals for these structures.)

absorption and emission. Some of the squaraine probes have very large two-photon cross-sections, as indicated in **Figure 10.12**.

Photostable probes have been mainly used in fluorescence microscopy—perhaps the most common use is to label antibodies (usually secondary antibodies) for visualization purposes. Occasionally, I have noticed that the structures of new probes are not always provided by the vendor—the vendor claiming that these are proprietary compounds and hence they cannot publish the structure! I would personally hesitate to use a probe of unknown (to me)

FIGURE 10.12 Left: Structure of squaraine-rotaxanes fluorophores for two-photon excitation. Right: Two-photon cross-sections of several squaraine and squaraine-rotaxanes fluorophores; dye names with a prefix of K8 denote squaraines, K9 prefixes denote squaraine-rotaxanes. (The author would like to thank Ewald Terpetschnig from Seta Biomedicals for these structures and spectra.)

structure—I would feel silly submitting a manuscript in which I write "The system was studied using a fluorescent compound of unknown structure."

Before leaving the topic of "photostability," I should address some of the questions that I have been asked over the years. Specifically, three of the most common questions beginners have asked are (1) Why does photobleaching happen? (2) Is it reversible? (3) How can I avoid it? Photobleaching generally occurs because a molecule in the excited state is more reactive than it is in the ground state. A practical aspect of photobleaching is that it limits the number of times a fluorophore can be excited and emit before it is destroyed. For example, fluorescein is estimated to be able to emit only on the order of 50,000 photons before being irreversibly photobleached. This aspect can be vexing for some experiments and helpful for others, as we shall see. A common excited state photo-induced process is reaction with oxygen—termed a photodynamic process. Photodynamic processes typically give rise to singlet oxygen species, which, in cells, can lead to significant cellular damage. These singlet oxygen radicals may also interact with and destroy ground-state fluorophores. To overcome oxygen-induced photobleaching, one can reduce the oxygen levels in the system—but, of course, this approach is not useful for work involving live cells and can be difficult even with *in vitro* systems.

One approach to reduce oxygen levels in cuvettes is to add a small amount of glucose and some glucose oxidase enzyme—and then sealing the cuvette against the atmosphere. Bubbling nitrogen or argon through the solution is another common approach. Another method, especially with cell systems, is to use antifading reagents, which are thought to act as radical scavengers. Of course the simplest approach, if possible, is to lower the excitation light intensity and to increase the detector sensitivity. In cuvette experiments, one can simply lower the lamp voltage or decrease the excitation slit widths. I commonly monitor the intensity of a sample's fluorescence with time, immediately after I place it in the instrument—specifically, I watch to see if the signal is stable or if there is evidence of photobleaching. If there appears to be bleaching, I will then decrease the intensity of the exciting light, either by using smaller excitation monochromator slits or by placing neutral density filters in the excitation path. As regards the reversibility of photobleaching, the process may not be reversible at the molecular level, but, in practice, in cuvette studies, one often finds that if the shutter is closed briefly then the signal recovers close to the starting signal. This recovery, of course, is due to the fact that only the excited fluorophores in the excitation light path were photobleached, and in a fluid solvent (such as aqueous solutions), these damaged molecules can diffuse out of the light path to be replaced by intact fluorophores. In fact, this recovery of fluorescence due to diffusion allowed development of the FRAP technique discussed in Chapter 9. Photobleaching can also be useful to reduce a cell's autofluorescence just prior to a fluorescence microscopy experiment. Finally, photobleaching of a potential acceptor in a FRET experiment can be used to verify FRET, as mentioned in Chapter 8. Reactions between fluorophores and molecular oxygen usually destroys the fluorophore and produces a free radical singlet oxygen species. This free radical can chemically modify other molecules in living cells and can be used for *photodynamic therapy*.

Labeling Proteins *In Vitro*
Noncovalent Probes

If your interest is to study some aspect of a protein using fluorescence spectroscopy, *in vitro*, and for various reasons you choose not to use the intrinsic protein fluorescence, then you must introduce a fluorophore into the system. In some cases, noncovalent probes may be useful. For example, 1,8-anilinonaphthalene sulfonate (ANS) and 4,4′-bis-1-anilinonaphthalene-8-sulfonate (bis-ANS) (**Figure 10.13**) have been used in a number of protein folding/ unfolding studies. The fluorescence properties of ANS were described in 1954 by Gregorio Weber and David Laurence, who noted that ANS was very weakly fluorescent in water but that its yield increased significantly, and its emission shifted to the blue, upon binding to bovine serum albumin (**Figure 10.14**) or to heat-denatured proteins. *Anecdote alert! The emission properties of ANS provide one of my favorite handlamp demonstrations of fluorescence— one which I highly recommend to anyone teaching an introductory class on*

FIGURE 10.13 Structures of ANS and bis-ANS.

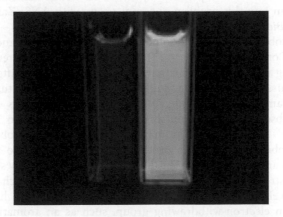

FIGURE 10.14 Solution of ANS in PBS (left) and the same concentration of ANS in PBS after addition of bovine serum albumin (right). Solutions are illuminated using a UV handlamp set for the long wavelength (366 nm).

fluorescence. One simply takes two large test tubes, one containing ANS in water, the other containing BSA in buffer. The exact concentrations are not so important—as long as there is a reasonable amount of probe and protein. Using a UV handlamp to illuminate the samples, one demonstrates that the ANS/water solution exhibits a very weak, yellowish fluorescence, while the BSA exhibits no fluorescence (there is sometimes a weak blue fluorescence from impurities, but it is usually negligible). With the lights out, you then pour the contents of one tube into the other (either way) and the result is a huge increase in fluorescence and a dramatic blue shift, that is, from weak yellow to very bright sky blue. This demonstration never fails to elicit "Oohs!" and "Aahs!" from the audience. It is interesting to note that even today, more than 50 years after the first report, ANS is still being used in protein studies, quite often as an indicator of the "molten globular" state. Bis-ANS, also introduced by Weber, boasts many of the properties of ANS and has also been shown to bind to many nucleotide binding sites on proteins. *Anecdote Alert! It is interesting and instructive to relate how Gregorio Weber first discovered bis-ANS. In Weber's laboratory, it was common to soak cuvettes in nitric acid solutions as part of the cleaning process. Apparently, in a few cases, even the exhaustive rinsing process failed to remove a small amount of nitrous ion, which was able to leach out of the walls of the cuvette during experiments on ANS binding to BSA. Weber and his postdoc at that time, Carl-Gustaf Rosen, noticed that the "ANS" in some cuvettes was binding much more tightly, and had different spectral properties, than they normally observed (Rosen and Weber, 1969, Biochemistry, 8: 3915). Further investigation tracked down the fact that the nitrous ions and BSA were able to catalyze formation of an ANS dimer, which they named bis-ANS. This discovery illustrates how serendipity can follow when one pays great attention to detail and gives in to curiosity.* Bis-ANS has also found use in protein unfolding/folding studies, although caution is warranted here since its fluorescence will increase in 1–2 M guanidinium chloride even without protein present (but not in urea), presumably due to molecular association with the guanidinium ion; this problem does not occur with ANS (Zakharov et al. 2011 *Anal. Biochem.* 416: 126).

Other probes, which bind to some proteins noncovalently, include thioflavin T and Congo Red (**Figure 10.15**), which have become extremely popular for following amyloid fiber formation. Thioflavin T exhibits shifts in its excitation and emission spectra and a significant increase in its quantum yield upon binding to amyloid-type protein. When Congo Red interacts with amyloid fibrils, its absorption maximum shifts from about 490 nm to 540 nm. Nile Red (**Figure 10.15**), although primarily known as a lipid probe, also binds to hydrophobic regions of some proteins and has been used to study protein aggregation and denaturation. The fluorescence yields of many such extrinsic probes are strongly influenced by the rate of intramolecular charge transfer, in which an electron is transferred from an electron donor, such as an amino group, to an electron-withdrawing group, such as an aromatic moiety. In the mechanism known as "twisted intramolecular charge transfer" or TICT, an alteration (such as rotation around a bond) in the molecule must occur for electron transfer to take place. Probes which demonstrate TICT are also

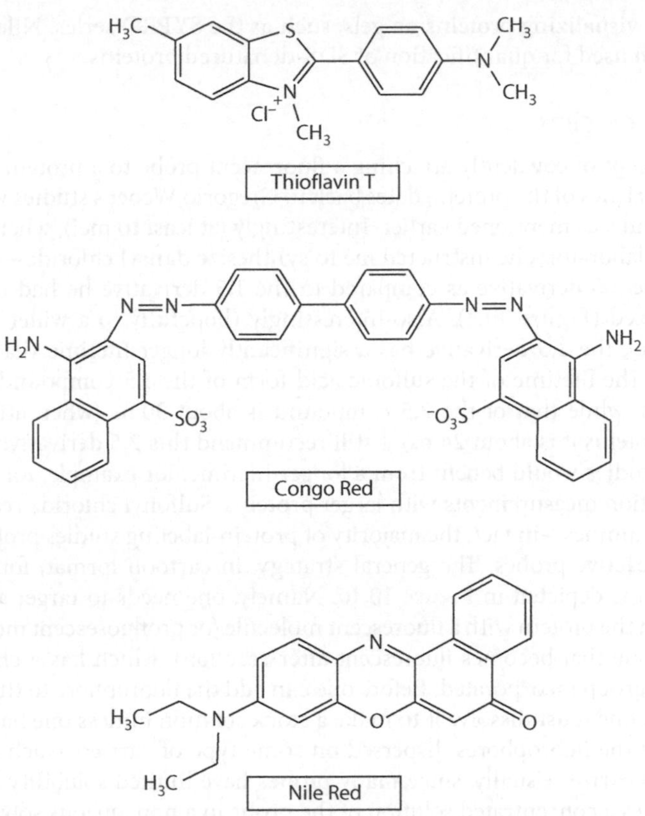

FIGURE 10.15 Structures of thioflavin T, Congo Red and Nile Red.

known as "molecular rotors" and have been used to sense solvent viscosities. Hydrogen bonding of a fluorophore with solvent molecules also can influence the spectrum and yield.

There are many proteins which specifically bind ligands that are intrinsically fluorescent or which can be rendered fluorescent, for example, NADH or FAD binding dehydrogenases (the emission properties of NADH, FAD, and some porphyrins will be discussed later in this chapter). Nucleotide binding proteins may also bind fluorescent nucleotide analogs—which also will be discussed later in this chapter. Specialized binding cases abound, of course, and many enzyme systems have been studied using fluorescent substrate/product/inhibitor analogs.

Protein quantification is usually accomplished using spectroscopic methods, such as the Lowry or Bradford assays, yet fluorescent assays are being increasingly used due to their high sensitivities. One of the more popular of these is the NanoOrange assay, which utilizes a merocyanine-based fluorophore, which, though poorly fluorescent in water, increases its quantum yield dramatically upon binding to heat-denatured proteins. There are also numerous fluorescent dyes, which interact strongly with SDS denatured proteins, used for visualizing proteins on gels, such as the SYPRO series. Nile Red has also been used for quantification of SDS-denatured proteins.

Covalent Probes

The concept of covalently attaching a fluorescent probe to a protein, to study characteristics of the protein, dates back to Gregorio Weber's studies with dansyl chloride, as mentioned earlier. Interestingly (at least to me!), when I joined Weber's laboratory, he instructed me to synthesize dansyl chloride—except to make the 2,5 derivative as compared to the 1,5 derivative he had originally synthesized (**Figure 10.4**). Also interestingly (hopefully to a wider audience this time), the 2,5 derivative has a significantly longer lifetime than the 1,5 version. The lifetime of the sulfonic acid form of the 1,5 compound is about 10–13 ns, while that of the 2,5 compound is about 30 ns (when attached to some proteins it is about 24 ns). I still recommend this 2,5 derivative to those whose studies would benefit from a longer lifetime, for example, for use with polarization measurements with larger proteins. Sulfonyl chloride reacts with primary amines—in fact, the majority of protein-labeling studies probably use amine-reactive probes. The general strategy, in cartoon format, for labeling a protein is depicted in **Figure 10.16**. Namely, one needs to target a reactive group on the protein with a fluorescent molecule (or profluorescent molecule— that is, one that becomes fluorescent after reaction), which has a chemically reactive group incorporated. Before one can add the fluorophore to the protein solution, one must dissolve it to make a stock solution (unless one has already obtained the fluorophores dispersed on some type of carrier—such as diatomaceous earth). Usually, since many probes have limited solubility in water, one makes a concentrated solution of the probe in a nonaqueous solvent, such as DMSO or DMF. Of course, one should check that a small amount of organic solvent will not adversely affect the protein. If the probe is soluble in aqueous solution, then by all means just use a buffer. Concentrated probe solutions are used so that one can keep the final concentration of organic solvent as low as possible (e.g., 1–3%) and also to avoid diluting the protein. A final note: back when the purity of commercial probes was not always assured, we typically ran

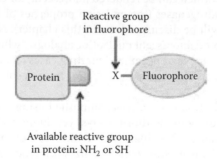

Reactive group
in fluorophore

Protein   X — Fluorophore

Available reactive group
in protein: NH_2 or SH

FIGURE 10.16 General approach to covalent modification of proteins with fluorescent probes.

thin-layer chromatography to check for contaminants. The widespread availability of HPLC or even mass spectrometry makes it easier to run such checks.

Amine Reactive Probes

As stated above, primary amine groups are probably the most targeted moieties in fluorescent labeling reactions. The principal reactive groups utilized to attach fluorophores to primary amines are (1) sulfonyl chlorides, (2) isothiocyantes, and (3) succinimidyl esters. Other chemistries are available (e.g., aldehydes react with amines to give a Shiff base, which must then be reduced, typically with borohydride), but these three are by far the most popular. The structures of these reactive groups and their products with primary amines are shown in **Figure 10.17**.

An important factor to consider when attempting to label primary amines is that the reactions target the $-NH_2$ form, not the $-NH_3^+$ form. Hence, the rate of the reaction is definitely pH dependent. Since the ε-amine moieties of most lysine groups in proteins have a pKa around 10.0–10.2, it is usually considered best to run the protein labeling reaction at basic pH values. How high a pH should you use? That depends first and foremost on what pH values your protein will tolerate. The optimal conditions for labeling a particular protein must be worked out for each individual case, that is, probe concentration, protein concentration, pH, reaction time, temperature, and so on. Typically, I start with pH 8.5, on ice for 1 h, using a buffer in which the protein is stable, with a probe to protein ratio of ~20–30. *But I emphasize that these conditions are not written in stone.* After determining the extent of labeling (to be discussed later), I may find that I need to adjust the conditions to get more or less probe incorporated. For example, if I am aiming for a final labeling level of one probe per protein and my first attempt gave only 0.1 probe per protein, then I may increase the time of the reaction or increase the probe to protein ratio or even the pH. I should comment that sulfonyl chlorides are notoriously unstable in aqueous solutions at elevated pH levels and so one may lose much of the labeling reagent during the incubation period—this fact has led some to disperse

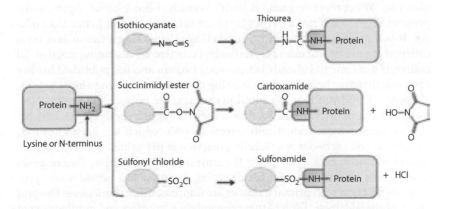

FIGURE 10.17 Reactions of amine reactive fluorescence probes with proteins.

the fluorophores on diatomaceous earth to protect them somewhat from hydrolysis by water. However, the final conjugate formed—sulfonamide—is very stable (although during long periods in storage some hydrolysis of such sulfonamide bonds has been observed). *Anecdote alert! When I was a graduate student, I was labeling mitochondrial malate dehydrogenase with FITC using a phosphate buffer of around pH 8.5. After separation of labeled protein from unreacted FITC, I began polarization studies—my goal was to study the protein's dimer to monomer dissociation. I noticed, though, that the polarization of the stock, that is, concentrated labeled protein solution, was dropping significantly over time. I dialyzed the sample against phosphate buffer and the polarization increased—suggesting that some free probe had been present but dialyzed away. But with time, the polarization of this dialyzed sample also began to decrease. I finally dialyzed the sample against tris buffer and the polarization of the dialyzed sample remained constant. My belief was that there was a very reactive tyrosine residue, which reacted with the FITC, but that this bond was not stable and had hydrolyzed over time. I further believed that dialysis against tris buffer, with its amine groups, facilitated the exchange of these unstable linkages with the solvent amines, hence removing unstable adducts but not affecting the FITC-labeled lysines. This strategy was tried by my friend Greg Reinhart, who mentioned that his label, pyrenebutyric acid, also appeared to be coming off of his labeled protein. Again, dialysis against tris buffer led to a stable adduct. I do not know how widespread this phenomenon of unstable labeling is, but I mention this story in the off chance that it may prove useful to some reader.*

Some proteins label much faster than others. For example, reaction of FITC with BSA proceeds very rapidly at pH 8, for example, 1 or 2 min reaction at room temperature is enough to get several fluoresceins covalently bound. Porcine mitochondrial malate dehydrogenase, however, under similar conditions will hardly react at all—it may take 24 h to get an average of one fluorescein bound. I should note that companies (such as Invitrogen) often provide explicit instructions for reacting proteins with their fluorescent reagents. One should start by following the specified conditions—but don't be shy about altering these conditions if you do not get the result you want! Readers may also ask, "What reactive group is best?" Frankly, I don't know! Again, some proteins seem to tolerate a particular procedure and reagent better than others. If you are not happy with your labeling attempts, you can always try a different chemistry, for example, sulfhydryl reactive versus amine reactive. Of course, if someone has already labeled your protein and has published his/her protocol, that may be an excellent starting place. Before leaving amine groups, I should remind you that at neutral pH values, the only amine in the $-NH_2$ form is the amino-terminus. So, unless there is a lysine residue in an unusual environment, which dramatically lowers its pKa (which has been known to happen), if you carry out your labeling reaction at pH values of 7 or lower, you should be able to exclusively target the amino-terminal residue. Two reagents that used to be quite popular for detecting amino acids, peptides and proteins via reaction with primary amines are fluorescamine (tradename Fluram) and o-phthalaldehyde. Both of these reagents are colorless and nonfluorescent until they react with amines to form highly fluorescent adducts. The structure

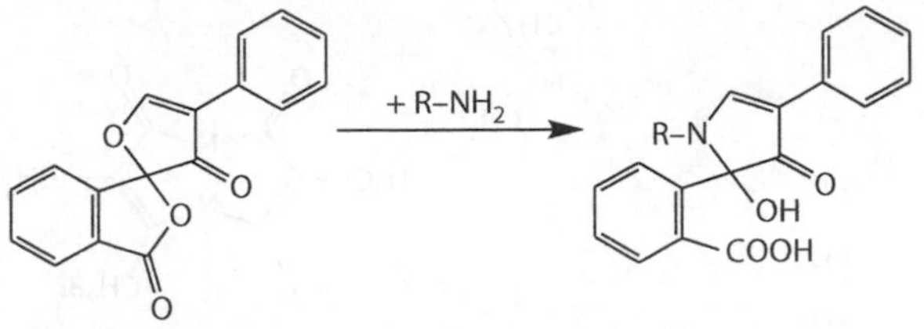

FIGURE 10.18   Structure and reaction of fluorescamine.

and reaction of fluorescamine are shown in **Figure 10.18**. Fluorescamine has also occasionally been used as a fluorescent probe in protein studies.

Thiol Reactive Probes

Cysteine residues are increasingly common targets for protein labeling. Perhaps, the main impetus behind this increasing popularity was the development of site-directed mutagenesis methods. The stunning advances in molecular biology have made it increasingly simple to insert or remove cysteine residues with one's protein *de jour*. One may then use fluorescent labels which target the thiol group of cysteine. As mentioned earlier, Gregorio Weber synthesized the first sulfhydryl reactive fluorescent probe, which was IAEDANS (depicted in **Figure 10.5**). The most popular sulfhydryl reactive groups presently utilized are alkyl halides and maleimides as shown in **Figure 10.19**. It goes without saying that if one keeps their target protein with reducing agents, such as dithiotreitol (DTT) or 2-mercaptoethanol, these must be removed during the labeling reaction. I should also mention acrylodan (**Figure 10.20**), the sulfhydryl reactive version of Prodan, which maintains its characteristic sensitivity to the polarity of its environment (discussed later in this chapter) and, hence, can be very useful for monitoring protein conformational changes. Another common thiol labeling reagent is monobromobimane (**Figure 10.20**).

Finally, I want to point out an obvious control experiment for those contemplating using any of these thiol probes. Namely, if one has inserted a cysteine residue into a protein with the aim of labeling, in addition to running the

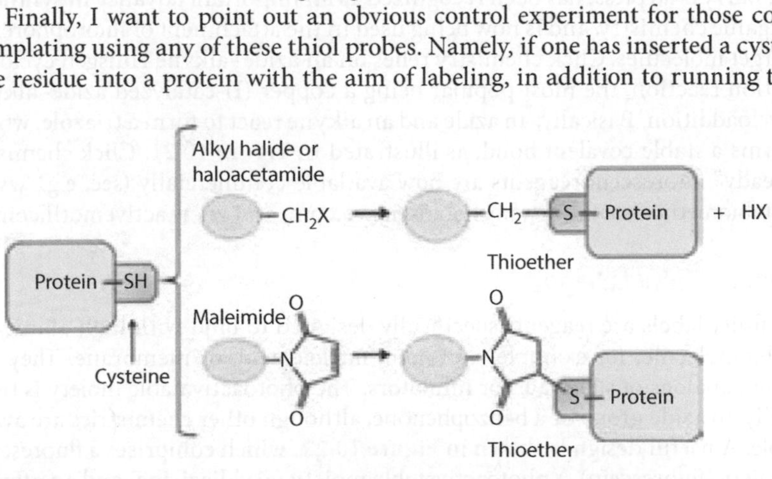

FIGURE 10.19   Reactions of sulfhydryl reactive fluorescence probes with proteins.

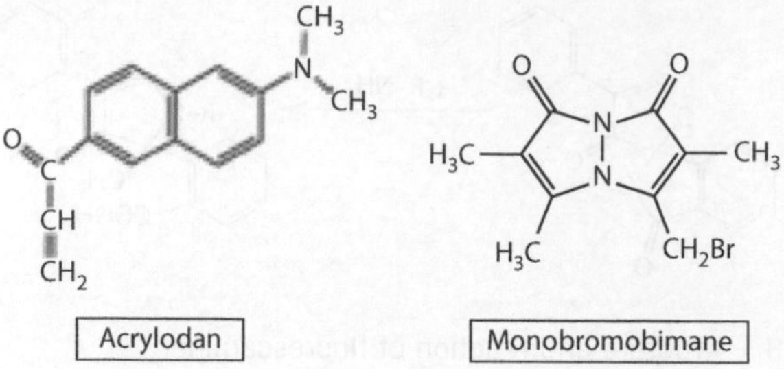

Acrylodan

Monobromobimane

FIGURE 10.20 Structures of acrylodan and monobromobimane.

labeling reaction with the cysteine-containing protein, one should also run the reaction with the wildtype protein, that is, the protein lacking the cysteine residue. The reason for this control is that these reagents are sulfhydryl *reactive*—not sulfhydryl *specific*! Alkyl halides and maleimides will react with other nucleophiles besides thiol groups. Thiol groups are strong nucleophiles at neutral pH, but if a lysine residue has its amine group in an unusual environment it can react—especially if one does not take care to keep the pH below 7. I have personally seen several cases in which the wildtype protein, without cysteines, was also labeled, meaning that in these systems some residue—probably lysine—in the control protein was in an environment that rendered it reactive with the thiol reagent. Texts which discuss the chemistry of protein reactive groups and diverse labeling reagents used for protein conjugation include Hermanson (1996) and Wong and Jameson (2012) (listed in Additional Reading).

Click Chemistry

In recent years, the method known as "click chemistry," developed largely by K. Barry Sharpless, has been recognized as an important advance in synthetic organic chemistry, and is now being used in the attachment of fluorophores to target molecules. Click chemistry relies on an azide–alkyne Huisgen cycloaddition reaction, the most popular being a copper (I)-catalyzed azide–alkyne cycloaddition. Basically, an azide and an alkyne react to form a triazole, which forms a stable covalent bond, as illustrated in **Figure 10.21**. Click chemistry "ready" fluorescent reagents are now available commercially (see, e.g., www.setabiomedicals.com, www.jenabioscience.com, and www.activemotif.com).

Photoaffinity Labeling

Affinity labels are reagents specifically designed to bind with high affinity to a biomolecule, for example, a protein, nucleic acid, or membrane. They are often analogs of substrates or inhibitors. The photoactivatable moiety is typically an azide group or a benzophenone, although other chemistries are available. A useful design is shown in **Figure 10.22**, which comprises a fluorescent moiety (fluorescein), a photoactivatable moiety (aryldiazirine) and an affinity ligand (a sidechain portion of aplyronine A). Such photoaffinity reactions are

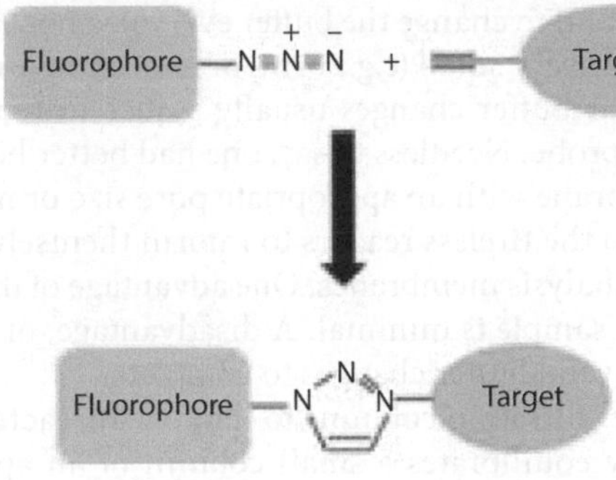

FIGURE 10.21 Example of a click chemistry reaction.

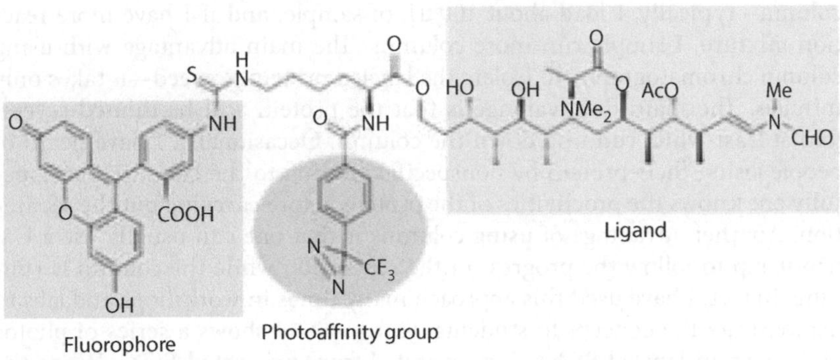

FIGURE 10.22 Example of a photoactivatable probe comprising a fluorescent moiety (fluorescein), a photoactivatable moiety (aryldiazirine) and an affinity ligand (a sidechain portion of aplyronine A). (Modified from T. Kuroda et al., 2006. *Bioconjugate Chem.* 17: 524.)

typically carried out using UV illumination, often from the 366 nm line of a standard UV handlamp.

Separation of Labeled Protein from Unreacted Probe

After the labeling reaction has run its allotted time, one must then separate the (hopefully) labeled protein from the unreacted probe. Some people recommend first stopping the reaction, for example, by adding cysteine to the labeling reaction in the case of sulfhydryl reactions or glycine in the case of amine reactions. Personally, I have rarely used this step, but I imagine there are cases where it is useful, for example, for more precise control of the extent of reaction. To remove unreacted probe, the most common approaches are either dialysis or column chromatography—each method has its advantages. For dialysis, one simply puts the reaction solution in an appropriate dialysis bag and places the bag in a large excess (typically 500–1000 fold) of the protein's favorite buffer. It is common to then place the buffer + dialysis bag + magnetic stirbar in a coldroom and to change the buffer every few hours. Since the reaction volume is typically very small (e.g., 1 mL or less) compared to the dialysis reservoir, three or four buffer changes usually suffice to remove free (unreacted or hydrolyzed) probe. Needless to say, one had better be sure that one is using a dialysis membrane with an appropriate pore size or molecular weight cutoff. I will leave it to the tireless readers to inform themselves of the proper care and handling of dialysis membranes. One advantage of dialysis is that the volume change of the sample is minimal. A disadvantage, of course, is that it can take hours and several buffer changes to complete.

The use of small gel-filtration columns to remove unreacted probe is fairly common. One simply equilibrates a small column of an appropriate material, such as a P-10 column (from GE Healthcare; which contains Sephadex G-25 Medium), with the protein's favorite buffer and then runs the reaction mixture over the column. Of course, one must guard against overloading the column—typically, I load about 100 µL of sample, and if I have more reaction mixture, I simple run more columns. The main advantage with using column chromatography to isolate the labeled protein is speed—it takes only minutes. The main disadvantage is that the protein will be diluted several fold at least while running down the column. Occasionally, I have heard of people losing their protein by nonspecific binding to the column, but hopefully one knows the proclivities of the protein before carrying out the separation. Another advantage of using columns is that one can usually use a UV handlamp to follow the progress of the separation while the column is running. In fact, I have used this approach many times in workshops and labs to demonstrate the concept to students; **Figure 10.23** shows a series of photos of fluorescein-labeled BSA being separated from unreacted FITC. When the labeled protein—which, of course, runs ahead of free dye on a size-exclusion resin—begins to exit the column, one can easily see the fluorescence from the drops, and, hence, can collect the labeled protein without even having to use a fraction collector. Another common method to remove free probe is to

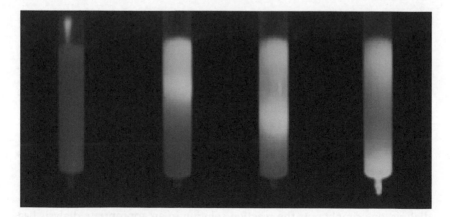

FIGURE 10.23 Time series photos of FITC-labeled BSA passing through a P-10 size-exclusion column, illuminated with a UV handlamp.

use a spin column. For example, spin columns loaded with gel permeation resins will allow rapid separation of proteins from small molecules, which are retained in the resin after spinning the tube with a benchtop microcentrifuge. Centrifugation filters, which are essentially small centrifuge tubes containing membranes with appropriate sized pores, provide another separation option.

Determination of the Extent of Labeling

A perennial question I am asked is, "Will the label disrupt the protein?" I usually reply, "I doubt it—but that's for you to find out!" In the vast majority of cases I am personally familiar with or have read about, there has been no effect, or minimal effect, of the fluorescent label on the biological activity. But exceptions occur, and each case much be checked individually. Of course, if one is labeling a cysteine residue and if it is known that there is an essential sulfhydryl in the active site, one may anticipate problems! One way to minimize labeling a residue necessary for activity is to carry out the labeling protocol in the presence of substrate or ligand.

To determine the average extent of labeling, one must ascertain the concentrations of both fluorophore and protein. Since virtually all extrinsic probes absorb light at longer wavelengths than proteins (excluding proteins with chromophores attached, such as porphyrins), the fluorophore concentration can usually be determined by measuring the optical density at a wavelength where the extinction coefficient is known. For example, I have used 71,000 M^{-1} cm^{-1} (488 nm) and 5500 M^{-1} cm^{-1} (351 nm) for the extinction coefficients for FITC and 1,5 AEDANS, respectively, bound to proteins (however, the absorption properties of fluorophores may change depending on their environments). Protein concentrations are usually determined colorimetrically using assays such as the Lowry, Bradford, Coomassie, or another, whatever is known to work best for your protein. If I only need a very rough estimate of the average labeling, in the case of FITC, I simply estimate the protein concentration by measuring the optical density at 280 nm, subtracting about one-fifth of the optical density at 480 nm, and then applying the Beer–Lambert law to the 280 nm OD, using the known protein extinction coefficient (which can be estimated from the amino acid composition as discussed in Chapter 2). This estimate makes the rough assumption that the fluorescein absorbance at 280 nm is about one-fifth of its absorption at 480 nm. I emphasize that this approach is only good for rough estimates, since, as mentioned, the fluorophore's absorption parameters may vary with its microenvironment. Some groups turn to mass spectrometry for definitive quantification of the extent of labeling.

Another point to consider is that if there are multiple possible labeling sites, which is likely when amine residues are targets, and if they have similar reactivities, then the distribution of labels among the protein population will follow a Poisson distribution, according to the equation:

$$P = \frac{n^{b}e^{-n}}{b!}$$

here P is the probability that a protein has a particular number of probes bound, b is the number of probes bound to a particular protein, and n is the average number of probes bound for the entire protein population. For example, if the average number of bound probes is 0.5 then 60.6% of the protein population has zero bound, 30.3% of the protein population has one bound, and 7.6% of the protein population has two bound. A table listing the percentage of the protein population with specific numbers of probes bound, as a function of the average probe to protein ratio, is given in **Table 10.1** (please note that these distributions assume random labeling).

Hence, if on average there is 1 probe per protein, then more than a third of the proteins will not have a probe attached, and more than a quarter will have more than one probe bound. These considerations can be important if bound probes are close enough to affect each other, for example, via FRET. A classic example, which comes up whenever I teach a workshop, is BSA labeled with FITC. If the average number bound is very low, such that most proteins have one or zero probes bound, then the observed polarization will be high (>0.4), reflecting the 4 ns lifetime and slow (~100 ns) rotational relaxation time of BSA (with limited local probe mobility). If there are more FITC molecules bound, though, which is easy to do given the large number of reactive lysine residues available, then the polarization can be much lower, for example, <0.3, due to depolarization via energy transfer between bound fluoresceins.

I should mention that overlabeling with xanthene (such as rhodamine) or cyanine dyes has been known to lead to the formation of dye dimers or excitons, which leads to decreased intensity. Nonfluorescent ground-state dimers were discussed briefly in Chapter 2, and absorption spectra for tetramethylrhodamine monomers and dimers were shown in Figure 2.10. In 1909, S.E. Sheppard wrote a seminal paper demonstrating that isocyanine showed a marked deviation from the Beer–Lambert law and specifically called attention to the appearance of the absorption spectra of isocyanine in organic and water solutions. He noted that in alcohol, isocyanine exhibited a maximum at 575 μm and a small shoulder at 535 μm, while in water the 535 μm

Table 10.1 Percentage of Protein Population with Specific Numbers of Probes Bound Presented as a Function of Average Probe to Protein Ratio

Average Number of Probes Bound (n)	Proteins with 0 Bound ($b=0$)	Proteins with 1 Bound ($b=1$)	Proteins with 2 Bound ($b=2$)	Proteins with 3 Bound ($b=3$)
0.5	60.6%	30.3%	7.6%	1.3%
1	36.8%	36.8%	18.4%	6.1%
2	13.5%	27.1%	27.1%	18.0%
3	5.0%	14.9%	22.4%	22.4%

band became a separate band, comparable in intensity to the 575 µm band. Sheppard even noted that increasing the temperature caused the aqueous absorption spectrum to revert more toward the alcohol spectrum. Thus began the formal study of dye dimers! Sheppard maintained his interest in dye molecules for many years—he became a prolific scientist at Kodak Laboratories, which, of course, had a great interest in the light absorption properties of dyes (these dyes were widely used in sensitizing the silver halide microcrystals in photographic films). For a classic early review by a pioneer in the field, read the 1942 paper by Sheppard listed in Additional Reading. It is probably safe to say, however, that Sheppard and his colleagues would not have imagined the current day wide range of applications for such dye aggregates in biochemistry, clinical chemistry, and molecular biology. An early connection between the absorption properties of certain dye aggregates and their fluorescence properties was illustrated by Edwin Jelley (also at Kodak) in 1936. Jelley gave the first description of the properties of what were later named "J-aggregates," in recognition of his work.

The exciton concept was introduced by J.I. Frenkel in 1931. In 1948, A.S. Davydov made important contributions to the theory, as did Michael Kasha in the 1960s. An in-depth consideration of excitons requires a quantum mechanical treatment—once again beyond the scope of this book. The arrangement of the monomers in the dimer complex will influence the extent of energy coupling in the system. When the monomers are arranged parallel to one another, the system is termed an H-dimer (larger assemblies are H-aggregates), and is essentially nonfluorescent; the head-to-tail arrangement is termed a J-dimer (J-aggregate), and the system can be fluorescent (Figure 10.24). Intermediate or inclined arrangements can give rise to weak fluorescence. Formation of such dye dimers can be a nuisance in many labeling studies, since the dimers can be nonfluorescent (or at least very weakly fluorescent). However, they can also be used as research tools, for example, to assess the proximity of protein domains (Hamman et al. (1996) *J. Biol. Chem.* 271: 7568) or as a spectroscopic signal for certain assays (Blackman et al. (2002) *Biochemistry* 41: 12244). Classic studies on xanthene dye dimer formations were carried out by Förster and König (1957), Selwyn and Steinfeld (1972), Arbeloa and Ojeda (1982), and Chambers et al. (1974) (references listed in Additional Reading).

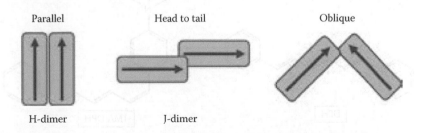

Parallel   Head to tail   Oblique

H-dimer   J-dimer

FIGURE 10.24 Depiction of molecular orientations of excitons.

Membrane Probes

Fluorescence probes have been used in studies of model and natural biological membranes for more than 40 years. To be useful as a membrane probe, a fluorophore must have a lipophilic aspect. One of the first membrane fluorescent probes was, in fact, our old friend ANS, which exhibits increasing fluorescence as the polarity of its environment decreases. ANS was used as early as 1969 to probe biological membranes. One of the first papers to use polarization to probe membrane-related systems was from Gregorio Weber's lab in 1971; this work introduced the use of perylene (**Figure 10.25**) in micelles, followed in 1973 by a paper on the use of perylene with model phospholipid systems. In 1974, Meir Shinitzky and Yechezke Barenholz introduced the probe diphenylhexatriene (DPH) (**Figure 10.25**) for the study of membrane systems. The fluorescence polarization of DPH was used extensively to study the physical

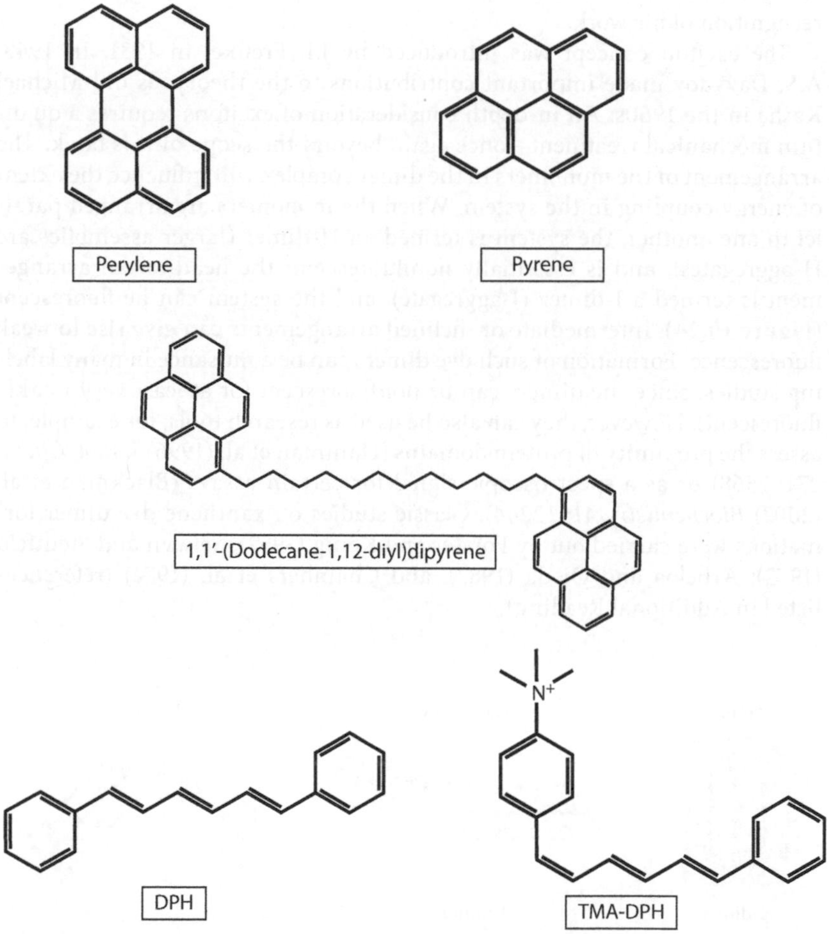

Perylene

Pyrene

1,1′-(Dodecane-1,12-diyl)dipyrene

DPH

TMA-DPH

FIGURE 10.25 Structures of various membrane probes.

state of model and natural membrane systems. DPH became by far the most popular membrane probe over the next two decades and many hundreds of papers were published using this probe—which in fact is still in use. DPH was shown to partition into the bilayer interior, and derivatives carrying charged groups, such as 1-[4-(trimethylamino) phenyl]-6-phenylhexa-1,3,5-triene, or TMA-DPH, a cationic analogue of DPH (**Figure 10.25**) were synthesized to anchor the fluorophore to the phospholipid side chain region. In this way, the probe does not readily translocate to other membranes in the cellular interior. In addition to fluorescence polarization studies, it has been shown that the lifetime characteristics of DPH depend on the physical state of the bilayer, presumably due to water penetration into the bilayer.

Pyrene (**Figure 10.25**) has also been widely used in membrane studies because of its ability to form excited dimers or *excimers*. Formation of an excimer requires that an excited pyrene molecule can form a complex with a ground state pyrene—which, of course, must occur during the lifetime of the excited state. Since pyrene excited states tend to be long (tens or hundreds of nanoseconds depending on the exact probe), the likelihood of such an encounter is reasonable if the concentration of the probe is sufficient and if the medium, in this case a biological membrane, is sufficiently fluid to ensure facile diffusion of the fluorophores. The excimer emission is red-shifted from the monomer emission so one simply has to determine the ratio of excimer to monomer emission to quantify the extent of excimer formation. An example of the use of pyrene excimers is shown in **Figure 10.26**. In this example, an aptamer probe for platelet-derived growth factor (PDGF) was synthesized with one pyrene molecule at each end. In the absence of the target protein, the aptamer adopts an open conformation such that each pyrene molecule is free to emit as a monomer (emission in the 350–440 nm range). When bound to the PDGF, however, the two pyrenes come into proximity and are able to emit as an excimer (emission above 440 nm). The increase in the excimer to monomer emission allows one to track the formation of the aptamer–PDGF complex. One can also use pyrene probes which contain two pyrene molecules connected by a linker. These types of dual probes have the advantage that both potential excimer partners are in the same molecule—hence, eliminating the requirement for the high probe concentrations necessary to ensure diffusional contact during the fluorescent lifetime.

The particular fluorescence technique utilized with a particular probe depends upon the photophysical properties of the probe as well as the type of information being sought. Of course, it also depends on whether an *in vitro* system is being probed, or if the system is a living cell. Let us first consider *in vitro* systems. For example, a very popular fluorophore used to probe membrane environments is Prodan, shown in **Figure 10.6** (and also Figure 5.6). This probe has a large dipole in the excited state due to the ability of the oxygen and nitrogen atoms to stabilize a negative and positive charge, respectively. It is this large dipole moment which renders Prodan sensitive to charges and dipoles in its surroundings. This excited state dipole, which exists only while the molecule is in the excited state, can act on its surroundings, especially if the surrounding molecules have dipole moments. Specifically, since the excited state dipole often has a different strength and direction (on

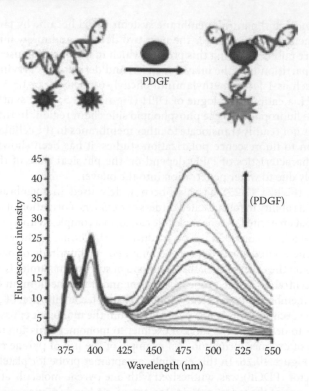

FIGURE 10.26 Example of the use of pyrene excimers. An aptamer probe for platelet-derived growth factor (PDGF) has one pyrene molecule at each end. In the absence of the target protein, the aptamer adopts an open conformation such that each pyrene molecule is free to emit as a monomer (emission in the 350–440 nm range). When bound to the PDGF, however, the two pyrenes come into proximity and are able to emit as an excimer (emission above 440 nm). (Adapted from C.J. Yang et al., 2005. *Proc. Natl. Acad. Sci. USA* 102: 17278.)

the nuclear framework), compared to the ground state dipole, the orientation of the solvent dipoles is energetically mismatched at the instant of the creation of the excited state. If the solvent molecules are able to move during the excited state lifetime, that is, if the solvent viscosity is not too high, then solvent reorientation can occur with a subsequent lowering of the excited state energy, as sketched roughly in **Figure 10.27**. As a consequence of these dipolar interactions, the energy of the excited state is decreased, which results in a red shift in the emission spectrum. The photo shown in **Figure 10.28** (also gracing the cover of this book), shows Prodan in glycerol at −80°C (top blue tube) and +60°C (bottom green tube) illuminated at 366 nm, using a UV handlamp. One can see the dramatic shift of the emission (from blue to green) as the glycerol dipoles are able to relax around the Prodan excited state dipole at the higher temperature. This process, termed *dipolar relaxation*, has been studied in detail in diverse systems including membranes and proteins

Franck–Condon state

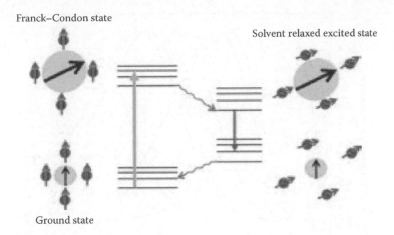

Solvent relaxed excited state

Ground state

FIGURE 10.27 Depiction of excited state solvent reorientation (dipolar relaxation).

(for protein studies see, e.g., the Macgregor and Weber and the Bismuto et al., papers listed under Additional Reading). Consequently, one may expect that the emission spectrum of a fluorophore such as Prodan will depend strongly on the polarity of the solvent. Spectra for Prodan in a series of solvents are shown in **Figure 10.29**.

The influence of solvent polarity on the absorption and emission properties of fluorophores, sometimes termed solvatochromic shifts, has been studied for many years. Different theories have been presented to treat solvent–fluorophore interactions, including those by E. Lippert, N. G. Bakhshiev, E. G. McRae, N. Mataga, A. Kawski, and G. Weber, among others. One goal of

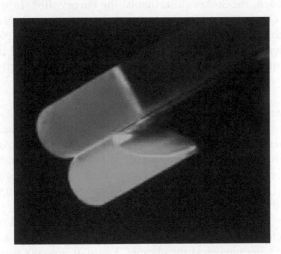

FIGURE 10.28 Prodan in glycerol at −80°C (top blue tube) and +60°C (bottom green tube) illuminated using a UV handlamp.

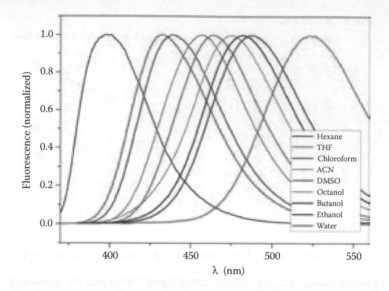

FIGURE 10.29 Emission spectra for Prodan in a series of solvents. (The author would like to thank Leonel Malacrida for these spectra.)

these theories is to estimate the excited state dipole moment. A detailed discussion of the various theories and their relative merits is well beyond the scope of this book (for extensive discussions see Reichardt (1994) *Chem. Rev.* 94: 2319 and Kucherak et al. (2012) *Phys. Chem. Chem. Phys.* 14: 2292). The most common treatment probably involves the so-called Lippert–Mataga plot. In this approach, the Stokes shift, that is, the energy shift (in wavenumbers) between the absorption and emission bands is assumed to be proportional to the dipole moment of the fluorophores upon excitation and is plotted against the orientation polarizability of the solvent. The orientational polarizability, Δf, of a solvent is related to its dielectric constant (ε) and refractive index (η):

$$\Delta f = [(\varepsilon - 1)/(2\varepsilon + 1)] - [(\eta^2 - 1)/(2\eta^2 + 1)]$$

These plots do not, however, explicitly consider the possibility of hydrogen bonding between the fluorophores and solvent, and consequently, solvents are sometimes separated into H-bonding and aprotic classes. An example of such a Lippert–Mataga plot is shown for Prodan in **Figure 10.30**. Sometimes, these plots are made by graphing the energy shift against the empirical polarity index $\Delta E_T(30)$, but as I have said, detailed discussion of this topic is beyond the scope of this book.

The probe Laurdan, which is Prodan with a lauric acid tail instead of the proprionic acid tail (**Figure 10.6**), has, in recent years, become a very popular probe for studying biological membranes. Laurdan presents very comparable fluorescence properties to Prodan but, due to its lauric acid tail, is more lipophilic and is thought to insert further into lipid bilayers than does Prodan. Both the excitation and emission spectra of Laurdan alter with the physical state of

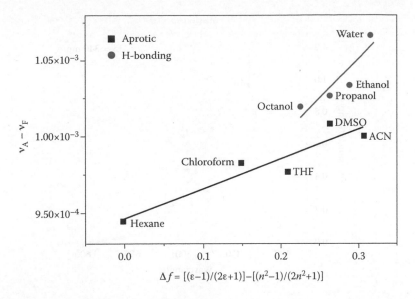

FIGURE 10.30 Example of a Lippert-Mataga plot for Prodan. (The author would like to thank Leonel Malacrida for this figure.)

the lipid bilayer, and this sensitivity forms the basis of the method used to follow alterations in the membrane's physical state. **Figure 10.31** illustrates the normalized emission spectra of Laurdan in a lipid bilayer, below and above its transition point. As usual with probes exhibiting spectral shifts, a form of ratiometric analysis is appropriate. Enrico Gratton and coworkers introduced the concept of the *Generalized Polarization* function, or GP, to serve this purpose (Parasassi et al. (1990) *Biophys. J.* 57: 1179). The concept underlying the

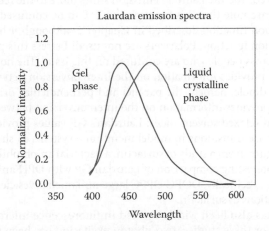

FIGURE 10.31 Depiction of normalized emission spectra for Laurdan in a lipid bilayer, below and above the phase transition point.

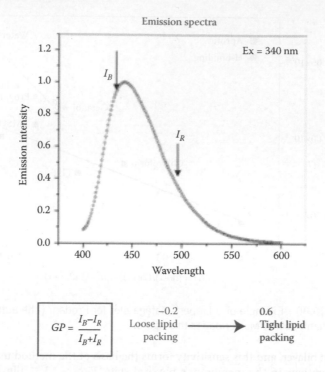

FIGURE 10.32 Illustration of the concept underlying the GP function.

GP function is illustrated in **Figure 10.32**. Typically, one determines the ratio of the emission intensity at 440 nm (the blue region) to that at 490 nm (the red region) and the GP function is calculated as illustrated. When lipid bilayers are more disordered, and the Laurdan emission shifts more to the red, the GP value decreases. Please note that this function is NOT to be confused with the normal polarization function described in Chapter 5, although it has the *form* of that polarization function. Polarizers are not used! I stress this point because I often come across people who are confused on this issue. The normal polarization function provides information on the interconversion between two states of an excited dipole, namely, the parallel and perpendicular orientations. The GP function provides information on the interconversion between the dipole's unrelaxed and relaxed solvent states. Laurdan's GP values provide a facile way to monitor phase transitions in model membrane systems as shown in **Figure 10.33**. This figure, from a study of surfactin, a bacterial amphiphilic lipopeptide, presents an interesting comparison of Laurdan GP with DPH anisotropy in the same systems, specifically, DOPC:DPPC large unilamellar vesicles with increasing concentrations of surfactin.

Laurdan has also been widely applied in fluorescence microscopy studies of live cells. For these studies, two-photon excitation has been the method of choice to minimize photobleaching of the Laurdan. By observing the emission alternatively through bandpass filters, passing the "blue" and "red" regions of

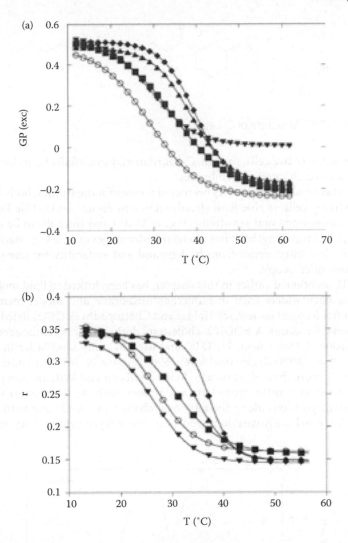

FIGURE 10.33 Example of the use of Laurdan's GP values to monitor phase transitions in a model membrane system. The figure (from M. Deleu et al., 2013. *Biochim. Biophys. Acta* 1828: 801) presents a comparison of (a) Laurdan GP with (b) DPH anisotropy in the same system, specifically, DOPC:DPPC large unilamaellar vesicles with increasing concentrations of surfactin.

the fluorescence, one can construct GP images, which provide information on the physical state of the cell membrane. In recent years, a modified form of Laurdan, named C-Laurdan, has been introduced, which has a carboxylic acid moiety attached (Kim et al. (2007) *ChemBioChem.* 8: 553) (**Figure 10.34**). Compared to Laurdan, this probe has higher water solubility, is less prone to aggregation in membranes, and localizes closer to the headgroup region of

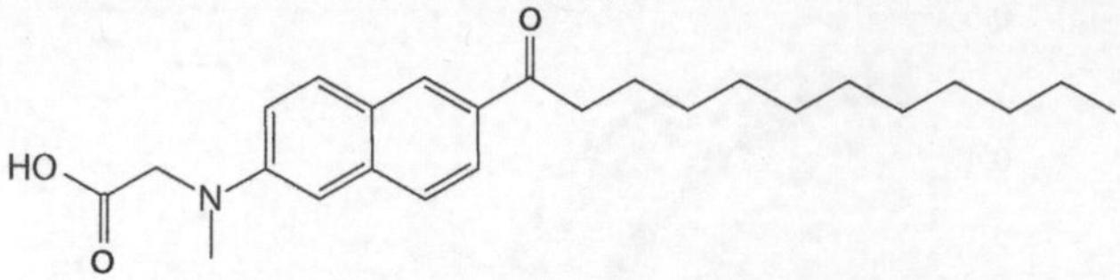

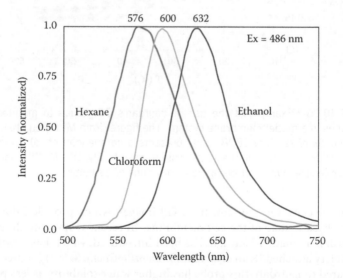

FIGURE 10.34 Structure of C-Laurdan.

lipid bilayers. In live cells, though, C-Laurdan may eventually be endocytosed and able to reach interior membranes.

Another lipid probe that has been used for membrane studies, both *in vitro* and in living cells, is Nile Red, already shown in **Figure 10.11**. Nile Red also displays environmental sensitivity (**Figure 10.35**) and since it can be excited using visible wavelengths, it has attendant advantages over UV excitation, for example, less interference from background and suitability for one photon excitation microscopy.

NBD, mentioned earlier in this chapter, has been linked to lipid molecules and has been widely used in numerous membrane and model membrane studies (for a recent review, see Haldar and Chattopadhyay (2013) listed under Additional Reading). A BODIPY-cholesterol derivative was synthesized by Li et al. (2006, *J. Org. Chem.* 71: 1718) and used to study lipid order in GUVs (Ariola et al. (2009) *Biophysical J.* 96: 2696) (**Figure 10.36**). As mentioned in Chapter 7, more than 30 years ago, Keith Thulborn and William Sawyer utilized a series of 9-anthrolyoxy fatty acid probes, with the fluorophore located at different positions along the fatty acid chain (2, 6, 9, 12) and with a fifth probe 9-methyl anthroate localized near the bilayer center. They utilized

FIGURE 10.35 Emission spectra of Nile Red in several solvents.

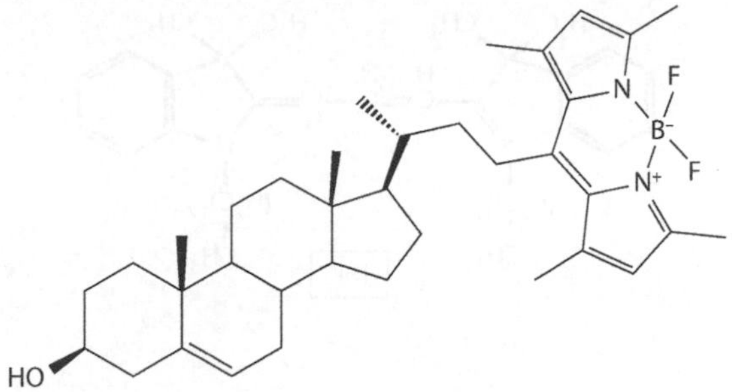

FIGURE 10.36 A BODIPY-cholesterol derivative.

fluorescent quenchers which partitioned to different depths in the bilayers to verify the positions of the fluorophores.

Additional Comments on Membrane Systems

The uses of fluorescent probes in live cell membranes typically fall into one of two categories, namely lipid/membrane staining or membrane dynamics. In this treatise, I shall not discuss membrane staining—first, because this area is well covered in the cell biology literature and second, because I know very little about it! Many of the lipid probes discussed above are well-suited for cuvette studies, but are more challenging to use with cells. Although, until the amazing advances in fluorescence microscopy in the last couple of decades, many researchers did (and some still do) carry out studies on cells using cuvettes and traditional spectrofluorimeters and spectrofluorometers. For example, DPH was added to cells in cuvettes and polarization measurements were used to study membrane fluidity. The essential drawback to cuvette approaches, however, was that it did not offer spatial information, that is, the signals observed arose from all parts of the cell. In the case of membranes, one could not be sure if the signal was mainly from the plasma membrane or if interior membranes contributed significantly. As fluorescence microscopy developed, probe chemistries and molecular biological methods also advanced and opportunites arose to obtain detailed spatial and temporal information from living cells. Some of these approaches were discussed in Chapter 9. Some of the most popular probes used for cell membrane studies are long-chain dialkylcarbocyanine dyes, such as DiI and DiO, shown in **Figure 10.37**. These probes are weakly fluorescent in aqueous environments but their yields increase substantially upon insertion into membranes; they are generally used to trace membrane structures and dynamics in live cells, for example, for cell–cell fusion studies. The spectral properties of these probes are largely independent of the lengths of the alkyl chains, and are instead determined by the heteroatoms in the terminal ring systems and the length of the connecting bridge. As mentioned earlier, Laurdan (and similar environmentally sensitive probes such as Nile Red) has become a popular probe for membrane studies on live cells. A wide-range of fluorophores have been attached to fatty acids and lipids. Voltage-sensitive

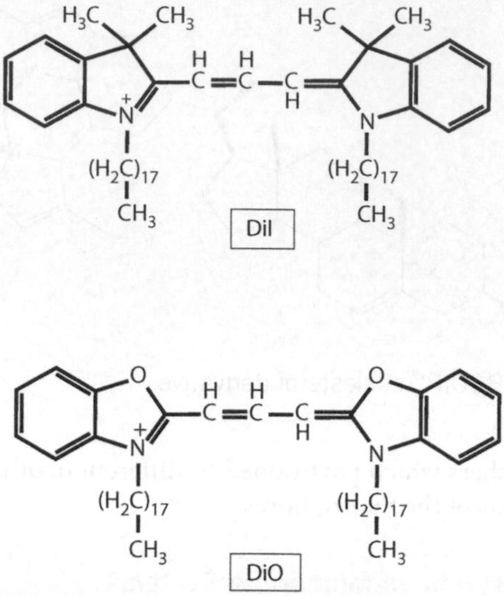

FIGURE 10.37 Dialkylcarbocyanines dyes, Dil and DiO.

dyes (discussed later) are also popular membrane probes. Some dyes will preferentially target specific cellular membranes. For example, MitoTracker Red CMXRos, from Life Technologies, stains mitochondria in live cells. One strategy for staining specific membranes, for example, the plasma membrane, is to attach a fluorescent dye covalently to an antibody specific for an antigen unique to that membrane. For example, wheat germ agglutinin (WGA) binds selectively to *N*-acetylglucosamine and *N*-acetylneuraminic (sialic) acid residues, and hence, labeling of WGA with a fluorophore will enable staining of membranes containing these sialic acid residues. Other dyes are available to preferentially target intracellular membranes. Those interested in these types of dyes should consult the copious literature.

Nucleic Acid Probes and Nucleotide/ Nucleoside Analogs

As mentioned earlier, the five natural nucleic acid bases (adenine, thymine, guanine, cytosine, and uracil) all have extremely low quantum yields, on the order of 0.5×10^{-4} to 3.0×10^{-4}. This low quantum yield for the components of DNA makes sense biologically, since one would not want genetic material to be able to hang around very long in an excited, and potentially reactive, state. The lifetimes associated with these emissions are on the order of picoseconds. A few naturally occurring, modified nucleosides, including 4-thioridine, 7-methylguanosine, N^6 acetylcytidine, and the Wye derivatives, have reasonable fluorescence and have been useful in studies of transferRNA. However, in order to apply fluorescence methods more generally to the study of nucleic

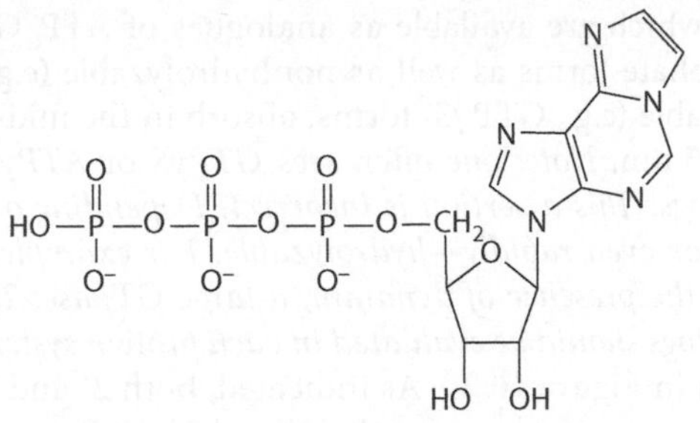

FIGURE 10.38 Structure of 1,N^6 ethenoadenosine, also known as ε-ATP.

acids, chemists began to synthesize modified bases. In 1970s, Nelson Leonard, at the University of Illinois at Urbana-Champaign, was one of the pioneers of this approach, which was not surprising since he was a friend and colleague of Gregorio Weber and their labs were not far apart (in fact, he was a member of my PhD committee)! Leonard and his coworkers inserted etheno bridges into purines and pyrimidines, which rendered them highly fluorescent. The 1,N^6 ethenoadenosine derivative, known as ε-ATP (**Figure 10.38**), is still a widely used nucleotide probe for diverse kinases and phosphatases.

Modification of the purine or pyrimidine ring structure, however, prevents binding of these analogs to some protein systems. Another strategy to synthesize fluorescent analogs thus focused on modification of the ribose-moiety. For example, TNP-GTP is a popular guanine nucleotide probe (**Figure 10.39**). Among the most popular nucleoside/nucleotide analogs in this group are the anthraniloyl and N-methylanthraniloyl derivatives of adenosine and guanosine, first synthesized and characterized by Toshiaki Hiratsuka in 1983.

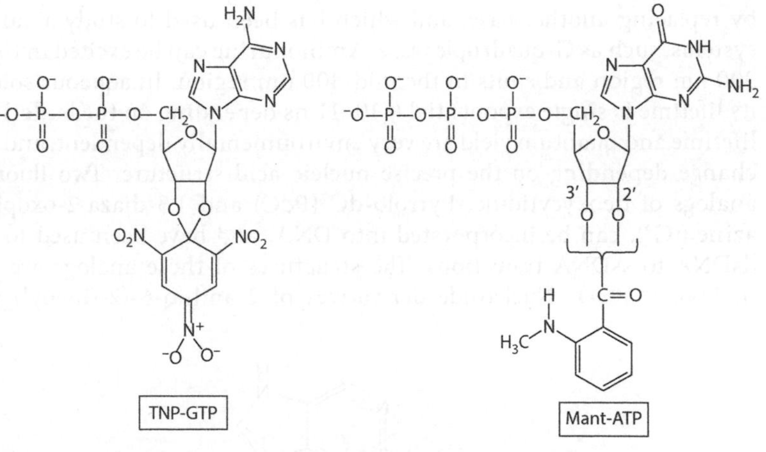

FIGURE 10.39 Structures of the nucleotide analogs, TNP-GTP and N-methylanthraniloyl-ATP (Mant-ATP). Both 2′ and 3′ isomers of Mant-ATP are shown.

These analogs, which are available as analogues of ATP, GTP, and their di- and monophosphate forms as well as nonhydrolyzable (e.g., GMPPNP) and slowly hydrolyzable (e.g., GTPγS) forms, absorb in the mid-300 nm and emit around 430–445 nm. *Note: one often sees GTPγS or ATPγS cited as nonhydrolyzable analogs. This assertion is incorrect! Depending on the system, they can be slowly—or even rapidly—hydrolyzable. For example, GTPγS is slowly hydrolyzable in the presence of dynamin, a large GTPase. The kinetic properties of these analogs should be evaluated in each protein system.* The mant-ATP analog is shown in **Figure 10.39**. As indicated, both 2′ and 3′ isomers can be made, but interconversion between the 2′ and 3′ positions occurs. Hence, to have a specific isomer, one must use either a 2′ or 3′ deoxy derivative. Often, but not always, the fluorescence yield of the mant derivatives increases upon binding to protein. When using mant probes in protein studies, one may be able to take advantage of the fact that tryptophan to mant energy transfer can occur, such that excitation of the probe-protein system at 280 nm leads to enhanced fluorescence yields of the mant moiety upon binding. Many ribose-modified nucleoside/nucleotide analogs are now available, with probes such as BODIPY, fluorescein, rhodamine, Alexas, Cy dyes, and others attached. One cautionary note—attachment of the fluorophore to either the 2′ or 3′ positions may not give identical outcomes. For example, when cyanine-modified ATP analogs interact with myosin, the change in fluorescence upon binding of the 2′ and 3′ isomers are distinctly different (Oiwa et al. (2003) *Biophys. J.* 84: 634). Readers interested in more information on the synthesis and applications of these types of ribose-modified probes can refer to the reviews by Jameson and Eccleston (1997), and Cremo (2003) listed in Additional Reading. Many fluorescent nucleosides/nucleotides are available commercially from Jena Bioscience (http://www.jenabioscience.com/).

One of the most popular analogs used to study DNA/RNA is 2-aminopurine (**Figure 10.40**), which can be incorporated into nucleic acid structures by replacing another base, and which has been used to study a variety of systems, such as G-quadruplexes. 2-Aminopurine can be excited in the mid-300 nm region and emits in the mid-400 nm region. In aqueous solutions, its lifetime is single exponential (~10–11 ns depending on the buffer) but its lifetime and quantum yield are very environmentally dependent, and hence, change depending on the precise nucleic acid structure. Two fluorescent analogs of deoxycytidine, Pyrrolo-dC (PdC) and 1,3-diaza-2-oxophenoxazine (tC°), can be incorporated into DNA, and have been used to follow dsDNA to ssDNA transitions. The structures of these analogs are shown in **Figure 10.41**. Nucleoside derivatives of 2-amino-6-(2-thienyl) purine

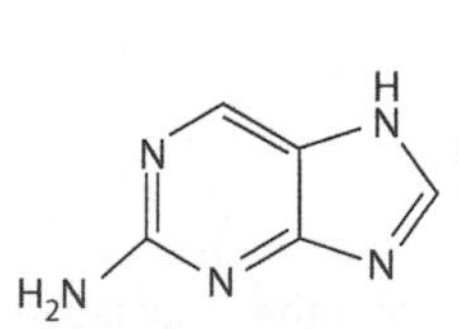

FIGURE 10.40 Structure of 2-aminopurine.

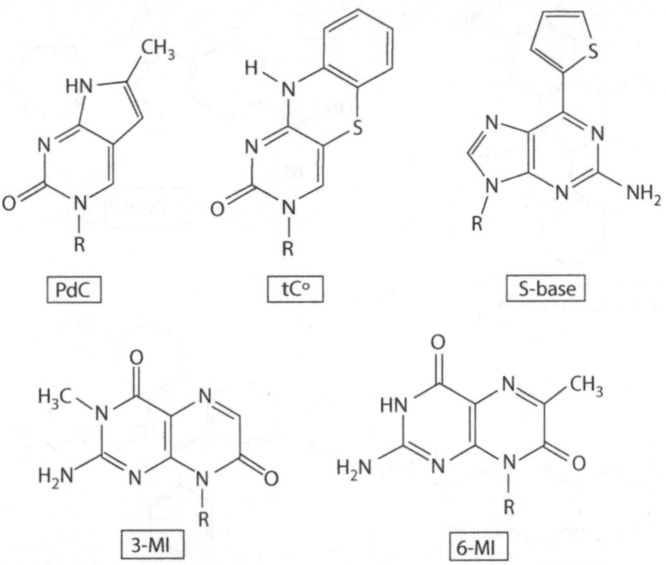

FIGURE 10.41 Structures of several nucleoside/nucleotide analogs.

(S-base) (**Figure 10.41**) are fluorescent and have been site-specifically incorporated into RNA and DNA. A fluorescent analog of guanine is 6-methyl isoxanthopterin (6-MI) (**Figure 10.41**), which has been used to monitor guanine base conformations at specific sites in RNA and DNA. 3-Methyl isoxanthopterin (3-MI) (**Figure 10.41**) has also been widely utilized in nucleic acid studies. *It is interesting to note that fluorescent pteridines were first isolated by Sir F. Gowland Hopkins from the wings of certain (Pieridae) butterflies in 1895 (F. G. Hopkins (1895) Phil. Trans. Roy. Soc., B, 186: 661). Indeed, in his paper describing these substances, he repeatedly remarked on their fluorescent properties!*

In recent years, many FRET studies, including single molecule studies, have been carried out on DNA and RNA structures using FRET donors and acceptors incorporated into the nucleic acid chain. Another important application of fluorescent nucleotide analogs occurs in DNA sequencing. The development of fluorescence approaches to DNA sequencing led directly to automation based on optical detection methods. In the dye-terminator approach to fluorescent-based DNA sequencing, each of the four dideoxynucleotide chain terminators is labeled with a different fluorophore and each emits at different wavelengths.

Fluorophores that intercalate into DNA or RNA structures are also popular. Perhaps the best known of this class of probes is ethidium bromide (**Figure 10.42**). Ethidium bromide (EtBr), which intercalates into double-stranded DNA or RNA, was originally used to treat trypanosomosis. Water is a highly effective quencher of EtBr fluorescence, which probably accounts for its highly increased yield when intercalated. The fluorescence lifetime of EtBr in aqueous solution is about 1.8 ns, but it increases to around 22 ns bound

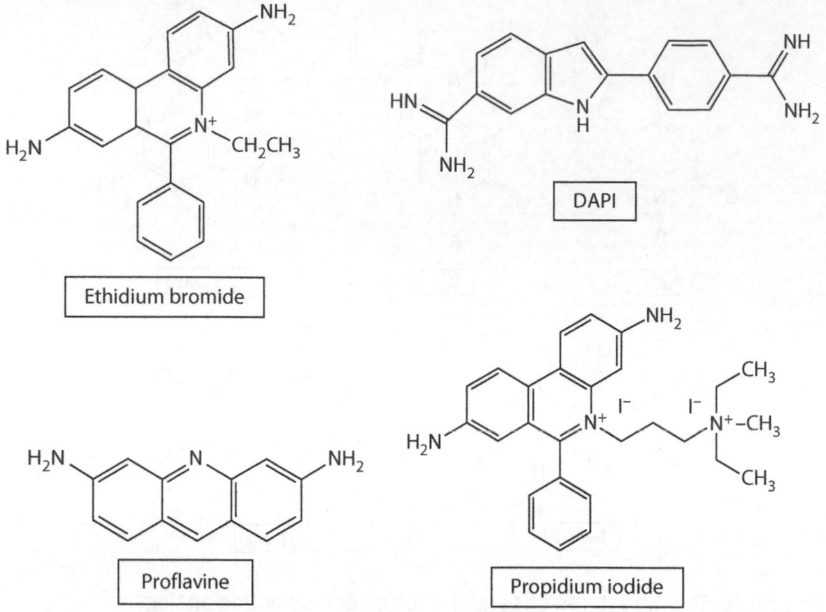

FIGURE 10.42 Structures of several fluorophores that intercalate into DNA or RNA structures.

to double-stranded DNA, and up to 27 ns bound to tRNA. Proflavine (also called diaminoacridine), propidium iodide, and DAPI are other heterocyclic structures, which can intercalate and exhibit significantly enhanced fluorescence (**Figure 10.42**). Interestingly, DAPI has also been shown to bind to the protein tubulin. A well-known series of dyes used to stain DNA in cells are the Hoechst stains. This family of probes is named for Hoechst AG, the German company which manufactured them (now a subsidiary of the Sanofi-Aventis pharmaceuticals group), and a probe's numerical designation (e.g., Hoechst 33258; **Figure 10.43**) indicates when that probe appeared in the series of compounds made by the company. Hoechst dyes are generally less toxic than DAPI and also more cell-permeable. These dyes generally absorb in the mid-300 nm range and emit in the mid- to high-400 nm range.

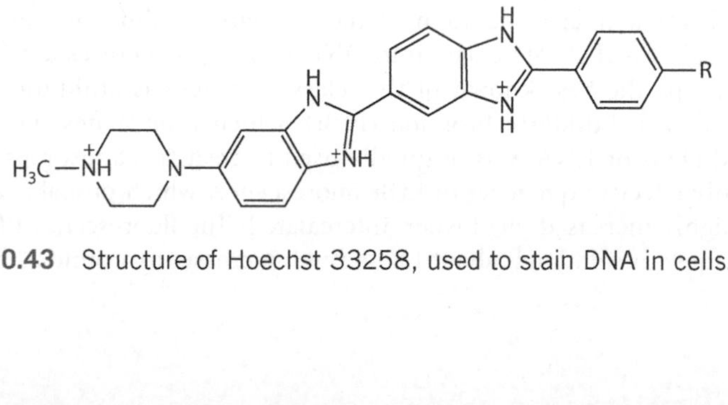

FIGURE 10.43 Structure of Hoechst 33258, used to stain DNA in cells.

Biosensors

Fluorescent biosensors are a class of specialized probes designed to detect and/ or quantify diverse atoms or molecules of biological relevance. Typically, biosensors contain one component, which binds to or reacts with the target, and another component that allows that reaction to be quantified. A huge variety of biosensors has been, and continues to be, designed and synthesized for diverse targets. These include relatively simple fluorophores, which can directly bind the target molecule as well as proteins or protein complexes, which have a fluorophore incorporated. In some cases, proteins have been engineered to enhance their ability to bind a target. It has become popular, even fashionable, to use biosensors composed of two fluorescent proteins, which can exhibit varying extents of FRET upon interaction with the target molecule. I shall mention only a few examples, but I caution to say that in the case of each target listed, other fluorescent assays have been developed.

1. Nitric oxide (NO) has been recognized as an important signaling molecule. In cells, nitric oxide synthase (NOS) catalyzes the conversion of arginine into citrulline plus NO. The NO can then diffuse through the cytoplasm and across cell membranes. The probe shown in **Figure 10.44**, 5,6-diaminofluorescein diacetate (DAF-2-DA), can diffuse into cell membrane and then be converted into the weakly fluorescent probe termed DAF-2. NO radicals can convert this molecule into the strongly fluorescent triazole DAF-2 T.

2. The classic example of a genetically encoded calcium biosensor that works inside cells, is the Cameleon sensor, developed by Roger Tsien and colleagues in 1997, depicted in **Figure 10.45**. As indicated, this biosensor is a recombinant protein which contains two fluorescent proteins (originally with BFP as donor and GFP as acceptor), which form a FRET donor/acceptor pair, the calcium binding protein calmodulin and the calmodulin binding peptide known as M13, which can bind to the calcium-bound form of calmodulin. The

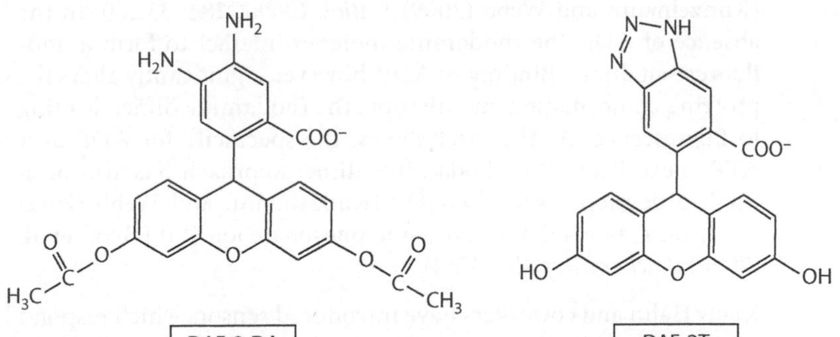

| DAF-2-DA | DAF-2T |

FIGURE 10.44 Structure of the diacetate form of DAF-2-DA, which is converted by NO radicals into the strongly fluorescent triazole, DAF-2 T.

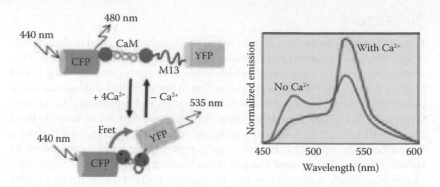

FIGURE 10.45 Left: Depiction of the Cameleon calcium sensor and its mode of action. Right: Sketch illustrating the change in emission spectra upon calcium binding to the Cameleon sensor.

extent of FRET changes (typically increases) when calcium binds to the calmodulin moiety, which in turn interacts with the M13 peptide. This interaction engenders either a change in distance or orientation (or both) of the donor/acceptor FRET pair, changing the FRET efficiency and providing a spectroscopic readout, namely the ratio of the acceptor-to-donor emission. Improved Cameleon sensors were described by Miyawaki et al. (1999) (*Proc. Natl. Acad. Sci. USA*, 96: 2135).

3. Martin Webb and his colleagues have taken advantage of the propensity of rhodamines to form nonfluorescent dimers (discussed earlier in this chapter) to develop several useful biosensors. For example, a biosensor to detect ADP is depicted in **Figure 10.46**. In this example, cysteine residues were engineered into an ADP binding protein (a mutagenesis-modified bacterial actin homologue, ParM), followed by attachment of rhodamine moieties (Kunzelmann and Webb (2009) *J. Biol. Chem.* 284: 33130). In the absence of ADP, the rhodamine moieties interact to form a nonfluorescent dimer. Binding of ADP, however, significantly alters the protein's conformation and disrupts the rhodamine dimer, leading to fluorescence. As the inset shows, the specificity for ADP over ATP is excellent. This rhodamine–dimer approach has also been used to develop assays for GDP (Kunzelmann and Webb (2011) *Biochem. J.* 440: 43) and inorganic phosphate ion (Pi) (Okoh et al. (2006) *Biochemistry* 45: 14764).

4. Klaus Hahn and coworkers have introduced sensors which respond to the nucleotide states of proteins such as Rac or Rho (Kraynov et al. (2000) *Science* 290: 333; Machacek et al. (2009) *Nature* 461: 99). These sensors are now termed FLARE biosensors, for *fluorescent activation reporter*. The salient aspect of this approach is that

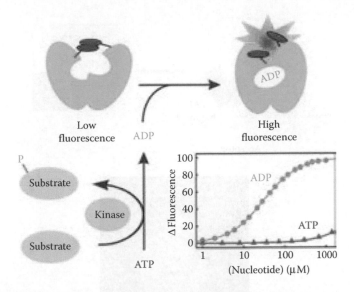

FIGURE 10.46 Illustration of the function of a biosensor, utilizing rhodamine dimers, designed to detect ADP. (The author would like to thank Martin Webb for this figure.)

it detects not just the presence of the protein of interest, but rather the active state, that is, GTP bound state, of that protein. Coupled with FRET imaging techniques, the method allows precise spatial and temporal localization of important cellular activities (**Figure 10.47**). The bottom part of **Figure 10.47** shows a cell image (a motile Swiss 3T3 fibroblast) in which FRET ratios have been converted into pseudo-colors, with the redder (warmer) colors indicating increased Rac1GTP levels.

Ion Probes

When ion probes are used in biological systems, they may be properly termed biosensors. Before the development of appropriate spectroscopic probes, ion-sensitive microelectrodes were commonly used. Now, a great variety of fluorescence probes are available to allow quantification of myriad ions in living cells and to provide precise spatial and temporal information. I cannot hope to cover all the different ion probes available, but I shall endeavor to mention several classes of probes and to give the general idea of their application.

Calcium probes act by chelating calcium ions and exhibiting a change in either their excitation or emission spectra, or a change in intensity or lifetime. All of these probes are based on linkage of a 1,2-bis(o-aminophenoxy) ethane-N,N,N',N'-tetraacetic acid (BAPTA) moiety to a fluorescence moiety. BAPTA has four carboxylic acid functional groups, which allows it to bind two calcium ions (**Figure 10.48**). Some of the earliest fluorescent calcium

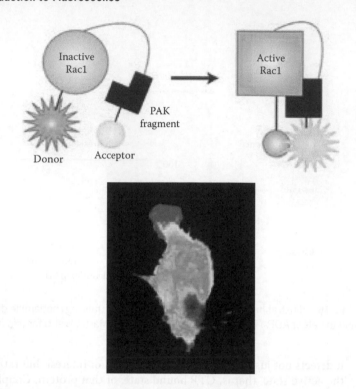

FIGURE 10.47 (Top) Diagram of a FLARE biosensor, which detects the GTP bound state of a target protein. PAK is a fragment of p21-activated kinase, which binds to and activates Rac1. Coupled with FRET imaging techniques, this biosensor allows precise spatial and temporal localization of important cellular activities. (Bottom) Cell image (a motile Swiss 3T3 fibroblast) in which FRET ratios have been converted into pseudo-colors, with the redder (warmer) colors indicating increased Rac1GTP levels. (The author would like to thank Betsy Clarke and Klaus Hahn for this figure. Information on many of these and other biosensors can be found on the Hahn Laboratory home page at hahnlab.com.)

probes were developed by Roger Tsien and his colleagues and include Indo-1, and Fura2, which had significant advantages over the popular probe Quin2. The structures of these probes are given in **Figure 10.48**. These probes allow for ratiometric detection; in other words, binding of calcium alters either the excitation or emission spectra of the attached fluorophores, which means that observation at two wavelength permits quantification of the extent of binding (**Figure 10.49**). Such ratiometric methods have a great advantage over direct, single wavelength intensity measurements, especially in live cell studies, since variations in probe concentration throughout a cell will affect intensities, but not ratio readings. Although these probes were effective, they could not be excited by the common laser line at 488 nm, which motivated the development of probes based on fluorescein and rhodamine structures. Fluo-3, Fluo-4, and Calcium Green are among the most popular of these probes (**Figure 10.50**).

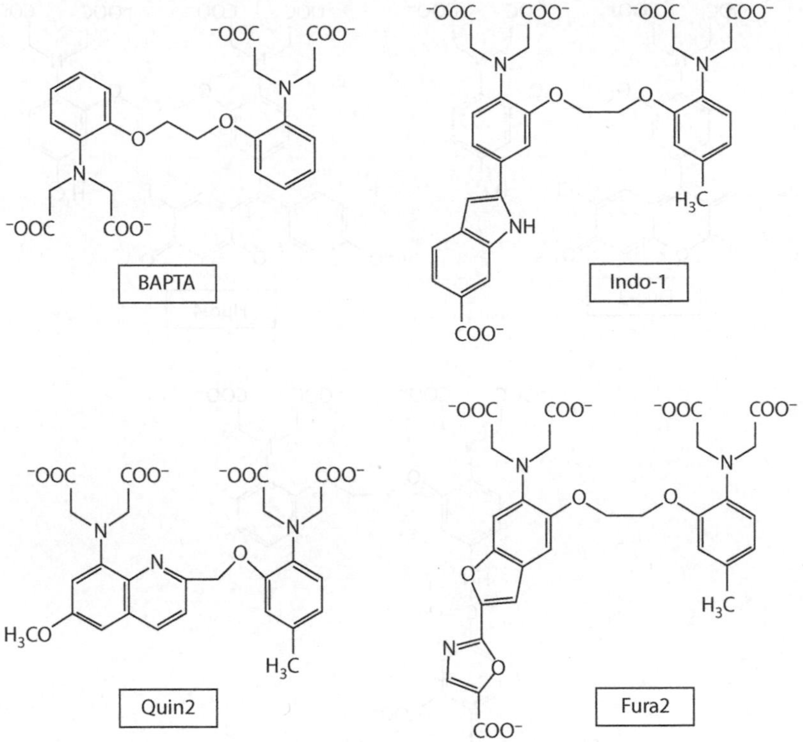

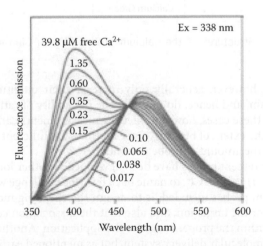

FIGURE 10.48 Structure of BAPTA, the core calcium binding module for numerous calcium probes, along with the structures of Indo-1, Quin2, and Fura2.

FIGURE 10.49 Emission spectra for Indo-1 as a function of free calcium. (The author would like to acknowledge Life Technologies for this image.)

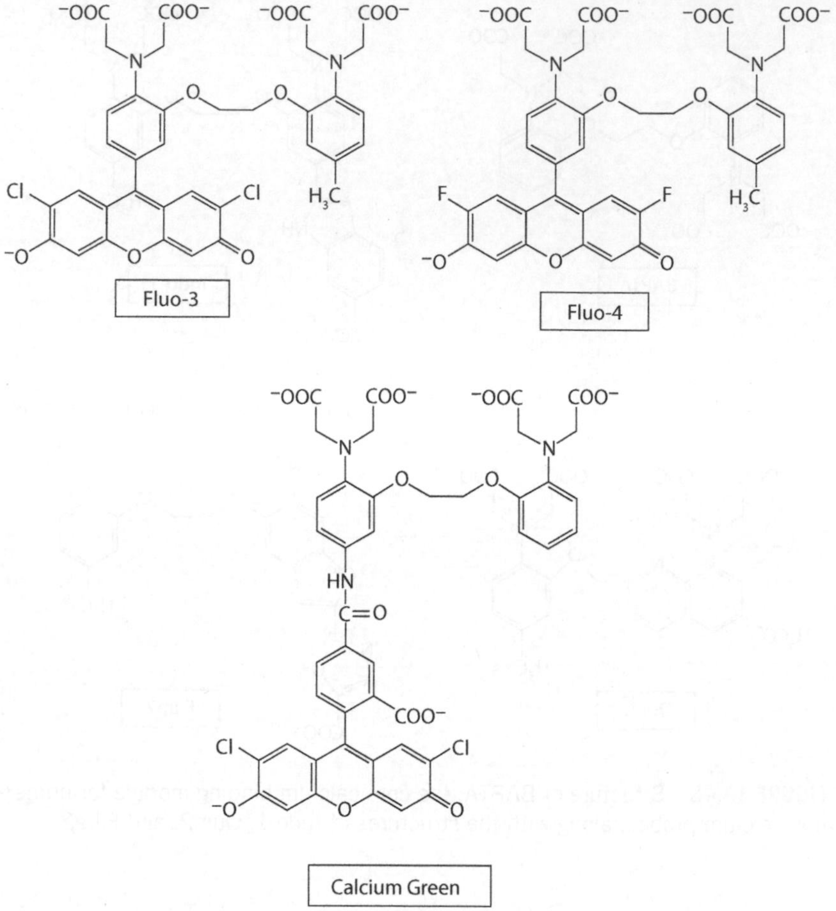

FIGURE 10.50 Structures of the calcium probes Fluo-3, Fluo-4, and Calcium Green.

These probes, however, generally only increase their quantum yield upon binding calcium, and hence, do not offer the possibility of ratiometric determinations. In these cases, however, lifetime measurements can often be used to determine the extent of binding, since lifetimes are inherent properties not dependent on the amount of probe present.

Similar fluorescent probes have been developed for other ions such as Na$^+$, Mg^{2+}, K$^+$, Zn^{2+}, Hg^{2+}, and Cl$^-$, to name but a few. The challenge with all of these probes, calcium probes included, is to design the chelating moiety such that it is selective for the target ion, and also that the dissociation constant for the target ion is within the proper range for the application. Another challenge is, of course, the molecular delivery system, but as mentioned earlier, most of the more popular probes are available in the ester form, which can be taken up by the cell, and which can then be processed by intracellular esterases.

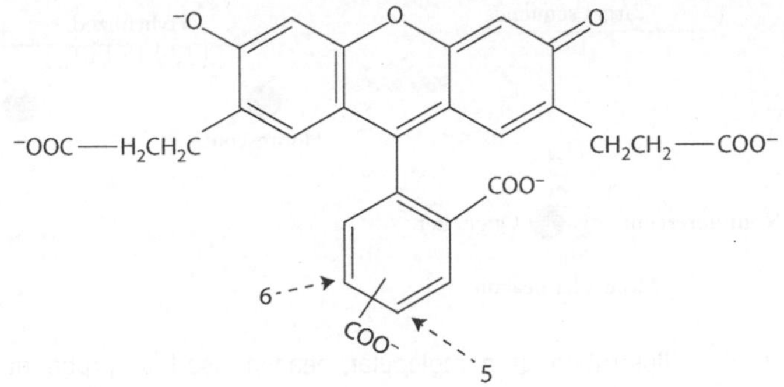

FIGURE 10.51 Structure of the pH probe BCECF.

pH Probes

Every chemist and biochemist understands that pH is very important for chemical reactions, both *in vitro* and in living cells. Litmus, a mixture of dyes found in different species of lichens, has been used for pH determinations for a really long time—since around 1300 AD! Litmus paper, developed in the early 1800s, and originally based on the chromophore 7-hydroxyphenoxazone, turns from blue to red in acid solutions. Electrochemists spent decades perfecting the standard laboratory pH meter, which became commercially available around 1935, thanks in large part to the efforts of Arnold Beckman. Determination of pH levels inside tissues and cells proved more daunting, however, since construction of microelectrodes small enough to probe live cells was clearly challenging. For this reason, optical pH sensors, and especially fluorescent pH sensors, have broad appeal. As mentioned in Chapter 1, Robert Boyle actually used the optical properties of *Lignum nephriticum* to judge the acid or alkaline nature of solutions. Fortunately, more useful fluorescent pH sensors are now available and are each characterized by a circumscribed pH range. One of the more common fluorescent pH probes is BCECF (2′,7′-bis-(2-carboxyethyl)-5-(and-6)-carboxyfluorescein), shown in **Figure 10.51**. One of the reasons for the popularity of this probe is that the pKa is 7.0, which is ideally matched to the normal range of cytoplasmic pH (~6.8–7.4). This probe is normally administered as the acetoxymethyl (AM) ester derivative, which is nonfluorescent and membrane permeant. Both the absorption and the emission of BCECF change, depending on the pH. Other fluorescent pH indicators, such as Oregon Green dyes (pKa ~4.7), LysoSensor Yellow/Blue DND-160 (pKa ~4.2), and SNARF (pKa ~7.5) cover a range of pH values.

Molecular Beacons

In 1996, Sangi Tyagi and Fred Kramer (*Nat. Biotechnol.* 14: 303) described a new approach to quantifying nucleic acid, which could be realized in real time and in solution, without the need for radioactivity or product separation. This approach

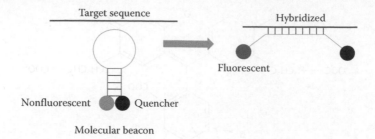

Target sequence

Hybridized

Fluorescent

Nonfluorescent   Quencher

Molecular beacon

FIGURE 10.52 Illustration of a molecular beacon used to probe nucleotide sequences.

was based on hairpin oligonucleotide probes they named *molecular beacons*. These molecules typically comprise a nucleotide sequence (e.g., 15–30 base pairs), which is complementary to the target nucleotide sequence, a stem region containing two nucleotide sequences complementary to each other that form base pairs, and fluorophores attached to the 5′ and 3′ ends (or a fluorophore attached to one end with a quencher attached to the other end). In the original design, a FRET pair was used to sense the opening of the hairpin nucleotide structure upon binding to a target sequence, as depicted in **Figure 10.52**. In other cases, quencher molecules have been used in place of fluorescent acceptors, which can give rise to a very significant increase in the fluorescence signal upon binding.

Voltage-Sensitive Dyes

Optical recording methods have become an important addition to the armamentarium of the neurobiologist. Voltage-sensitive dyes change their spectral properties in response to voltage changes, for example, in neuronal cells response to action potentials. Properly responding probes have the potential to gain information previously only available using electronic probes. The great advantage of optical methods over electrodes is, of course, that they can observe small structures, such as dendritic spines, which would be difficult to target even with microelectrodes. Voltage-sensitive probes appeared in the 1970s, but since then, many improvements in the response characteristics of these dyes have been made. For example, a class of fast-response dyes, known as charge-shift probes, based on (aminostyryl) pyridinium chromophores (also known as hemicyanine dyes), was introduced by Leslie Loew in 1985. An example of one of these probes, di-4-ANEPPS, is shown in **Figure 10.53**. More recently, Loew's lab showed that fluorination of such probes (**Figure 10.53**) can improve their spectral characteristics, and allow them to sense lipid order in model and cell membranes. Fluorescent proteins which can act as voltage-sensitive probes have also been developed, following the original fusion of GFP with the Shaker potassium channel (Siegel and Isacoff (1997) *Neuron* 18: 735). One of the latest in this family is the so-called ArcLight protein (Jin et al. (2012) *Neuron* 75: 779), which consists of the voltage-sensing domain of

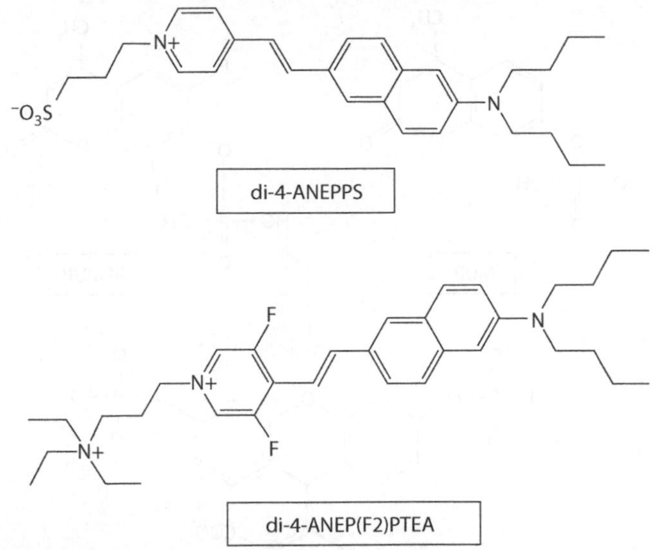

FIGURE 10.53 Structures of two voltage sensitive dyes. (Top) di-4-AminoNaphthenyl-Pyridinium-PropylSulphonate (di-4-ANEPPS). (Bottom) 4-[2-(6-Dibutylamino-naphthalen-2-yl)-vinyl]-1-(3-triethylammonio-propyl)-3,5-difluoro-pyridiniumdibromide (di-4-ANEP(F2)PTEA).

a voltage-sensitive phosphatase from *Ciona intestinalis*, and the FP known as super ecliptic pHluorin, carrying the point mutation A227D.

Miscellaneous Probes

Numerous enzyme assays have been developed which use the so-called "fluorogenic" substrates which are nonfluorescent as substrates but which give rise to fluorescent products. Examples of fluorogenic substrates have already been presented, namely fluorescamine (**Figure 10.19**) and 5,6-diaminofluorescein diacetate (**Figure 10.48**), or any similar nonfluorescent ester of a fluorophore that will become fluorescent after an esterase does its job. Other examples are the profluorogenic probes, 4-methylumbelliferyl phosphate (MUP) and 6,8-difluoro-4-methylumbelliferone DiFMUB (**Figure 10.54**), which can be used to detect phosphatase activities in high-throughput screening. Fluorescein diphosphate (**Figure 10.54**), which is colorless, can also be used to detect phosphatases, since hydrolysis of both phosphate esters creates fluorescein. Another class of long lifetime probes are based on transition metal ions (ions with an incomplete *d* sub-shell), such as ruthenium.

NADH, etc.

Perhaps a few details about the fluorescent properties of some of the intrinsically fluorescent molecules are warranted. The structures of some of these

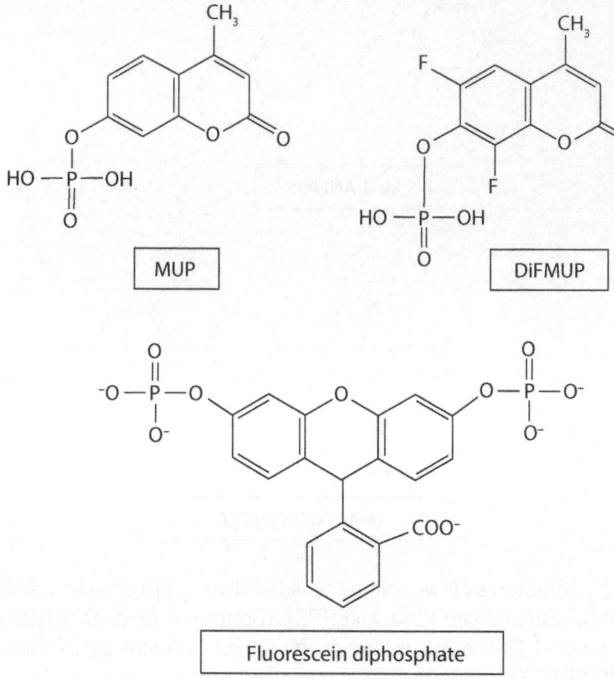

FIGURE 10.54 The profluorogenic probes, 4-methylumbelliferyl phosphate (MUP), 6,8-difluoro-4-methylumbelliferone (DiFMUB), and fluorescein diphosphate.

molecules were shown in **Figure 10.2**. NADH free in solution shows two major absorption peaks, near 260 nm and 340 nm, whereas NAD only shows the 260 nm peak. The emission of NADH, free in solution, is near 440 nm and its lifetime is short—around 0.4 ns. When bound to proteins, such as diverse dehydrogenases (e.g., lactate dehydrogenase, malate dehydrogenase, liver alcohol dehydrogenase, and others), the lifetime can increase significantly. In the case of lactate dehydrogenase, the NADH lifetime is about 4.5 ns, while in the case of the ternary complex of mitochondrial malate dehydrogenase, that is, with hydroxymalonate also bound, the NADH lifetime reaches 9.5 ns. FAD absorbs from the ultraviolet out to the visible region. The low-energy absorption band is around 500 nm and the emission maximum is near 530 nm; the lifetime of FAD in solution is near 4.5 ns. Porphyrins are, of course, important prosthetic groups in heme proteins. Although porphyrins bound with iron are nonfluorescent, those bound with some other metals, specifically closed-shell diamagnetic metals such as zinc, tin, and magnesium, do fluoresce, usually well into the red, that is, >600 nm. Of course, magnesium-bound porphyrins account for the red emission of chlorophylls.

Quantum Dots

Semiconductor quantum dots (QDs), with core diameters in the range of 1–10 nm, are very bright luminescent nanocrystals, which have found wide applications in biological imaging. The discovery of QDs is usually credited to A. I. Ekimov and A. A. Onushchenko, who published a report on quantum size effects in CuCl crystals in 1981. Although numerous group III–V, II–VI, or IV–VI elements have been used to fabricate QDs, the most popular QDs presently used seem to be cadmium based, such as CdSe or CdTe. The toxicity of cadmium has, however, in recent years, led the push for more biocompatible materials. QDs typically comprise hundreds to thousands of atoms and absorption of light leads to electron–hole recombination, which gives rise to luminescence. The QD core is surrounded by a shell, for example, ZnS; aqueous coatings can be added to facilitate solubility in biological media. The appeal of QDs is due to their high molar extinction coefficients ($10^5–10^7$ $M^{-1}cm^{-1}$), their quantum yields (up to 90%), their excellent photostability, the fact they absorb over a wide wavelength range and that the absorption and emission wavelengths depend upon the particle size (a quantum confinement phenomenon) and hence can be "tuned" for the task at hand (**Figure 10.55**). An example of

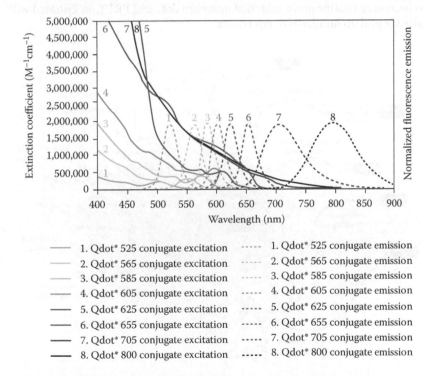

—— 1. Qdot* 525 conjugate excitation	- - - - 1. Qdot* 525 conjugate emission
—— 2. Qdot* 565 conjugate excitation	- - - - 2. Qdot* 565 conjugate emission
—— 3. Qdot* 585 conjugate excitation	- - - - 3. Qdot* 585 conjugate emission
—— 4. Qdot* 605 conjugate excitation	- - - - 4. Qdot* 605 conjugate emission
—— 5. Qdot* 625 conjugate excitation	- - - - 5. Qdot* 625 conjugate emission
—— 6. Qdot* 655 conjugate excitation	- - - - 6. Qdot* 655 conjugate emission
—— 7. Qdot* 705 conjugate excitation	- - - - 7. Qdot* 705 conjugate emission
—— 8. Qdot* 800 conjugate excitation	- - - - 8. Qdot* 800 conjugate emission

FIGURE 10.55 Absorption and emission spectra of a series of quantum dots. (The author would like to acknowledge Life Technologies for this figure.)

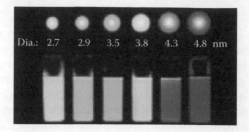

FIGURE 10.56 Illustration of the effect of core size upon Q-dot spectral properties. (The author would like to thank Russ Algar for this image reproduced with permission from E. Petryayeva et al., 2013. *Applied Spectros.* 67: 215.)

the effect of QD size on emission wavelength is shown in **Figure 10.56** from Petryayeva et al. (2013, *Appl. Spectros.* 67: 215), which provides an excellent review of the field. The shells of QDs can be functionalized to allow their conjugation with biomolecules, for example, proteins. Numerous functionalization strategies have been developed and some of these are illustrated in **Figure 10.57** from the Petryayeva et al. review. The use of QDs in FRET applications is becoming routine and a search of quantum dots and FRET on PubMed will already pull up hundreds of references.

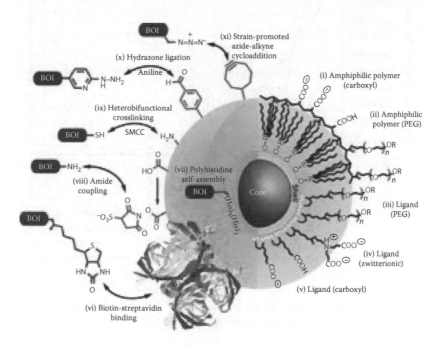

FIGURE 10.57 Illustration of functionalization strategies for Q-dots. (The author would like to thank Russ Algar for this image reproduced with permission from E. Petryayeva et al., 2013. *Applied Spectros.* 67: 215.)

Nanodiamonds

Nanodiamonds are not yet as popular as QDs but are finding increasing applications in the life sciences. The fluorescence from nanodiamond particles originates from defects, or color centers, in the crystalline matrix. Perhaps, the biggest advantage of fluorescent nanodiamonds (FNDs) over QDs is that FNDs are inherently nontoxic and biologically inert (although some recent studies have suggested that surface functionalized FNDs may show toxicity). The most common color center utilized to date appears to be the nitrogen-vacancy (NV) defect, responsible for red/near-infrared fluorescence. FNDs have excellent photostability and the FNDs characterized to date generally emit in the 550–800 nm range, although no doubt this range was already extended by the time this book appeared. FNDs are now finding applications in super resolution imaging studies. Needless to say, these highly technical advances are beyond the scope of this book. Interestingly, an economical way to fabricate FNDs is by the detonation of carbon-based explosives; these types of FND are also known as "detonation nanodiamonds." Surface functionalization of FNDs is still required, though, for conjugation with biomolecules such as proteins.

Fluorescent Proteins

Wow—where to start? Interest in fluorescent proteins has exploded during the past two decades. Any attempt to review the field of fluorescent proteins brings to mind the story of the man who drives a very old car into a gas station and asks for a fill up (this was, of course, in the days before self-service). After a short while, the attendant, who was attempting to fill up the car's gas tank, said to the driver, "Would you mind turning off the engine? You're getting ahead of me!" So, even though I will try to be current in my discussion, I'm sure that those making new recombinant fluorescent proteins are getting ahead of me!

I should first specify, though, that by "fluorescent proteins" I mean the class of proteins originally found in jellyfish and corals, as exemplified by the iconic green fluorescent protein (GFP), from *Aequorea victoria*. It is worth remembering that proteins with visible fluorescence had been known for a long time. In fact, Gregorio Weber's PhD thesis project was to understand the nature of the fluorescence from flavin and flavin-binding proteins, such as riboflavin. As mentioned, a great many dehydrogenases bind NADH and fluoresce in the mid-400 nm region (my thesis project concerned mitochondrial malate dehydrogenase). Other proteins which fluoresce in the visible region include phycobiliproteins (from cyanobacteria and eukaryotic algae), which are incredibly photostable and which possess extremely high extinction coefficients (some are higher than 2,000,000 M^{-1} cm^{-1}). The absorption and emission is due to their covalently attached tetrapyrole ring systems. Proteins associated with chlorophyll also fluoresce, of course.

The past decade has witnessed an explosion in the use of the family of naturally fluorescent proteins known as green fluorescent proteins or GFPs. GFP, a protein containing 128 amino acid residues, was originally isolated from the Pacific Northwest jellyfish *A. victoria*. In Chapter 9, a brief overview of the early work on GFP was given. The fluorescence moiety in GFP, contained within a β-sheet barrel, derives from autocatalytic and cyclization reactions involving three amino acid residues, namely Ser65, Tyr66, and Gly67 (**Figure 10.58**) (from Day and Davidson (2009) *Chem. Soc. Rev.* 38: 2887). This reaction is unusual and occurs only because of the unique environment offered by the overall GFP structure. The position of this chromophore in the protein structure is indicated in **Figure 10.59** (from Day and Davidson (2009) *Chem. Soc. Rev.* 38: 2887). The biological function of GFP is presumably to absorb a photon emitted by Aequorin, a protein originally isolated by Osamu Shimomura, which catalyzes a chemiluminescence in the presence of calcium. By absorbing the blue photon (~470 nm) from the Aequroin complex and reemitting a green photon (~520 nm), GFP shifts some of the emission from the jellyfish to longer wavelengths, which presumably confers a biological advantage. The green photon, compared to a blue photon, is less prone to scattering processes as it traverses the ocean environment and hence may convey information over longer distances. For example, release of this green photon in response to distress from a predator may attract a larger predator to the scene (which hopefully could distract the initial predator). In the 1990s,

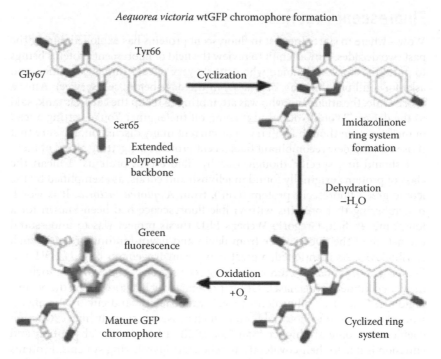

Aequorea victoria wtGFP chromophore formation

FIGURE 10.58 Illustration of the autocatalytic and cyclization reactions of the three amino acid residues which form the fluorophore in GFP, namely Ser65, Tyr66 and Gly67. (The author would like to thank Richard Day for this figure from R.N. Day and M.W. Davidson, 2009. *Chem. Soc. Rev.* 38: 2887.)

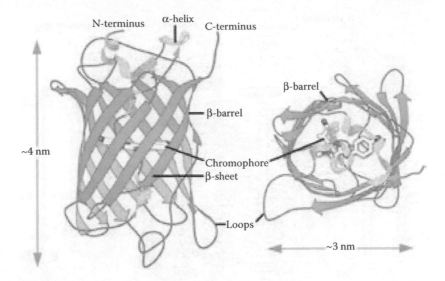

N-terminus   α-helix   C-terminus

β-barrel

β-barrel

~4 nm

Chromophore

β-sheet

Loops

~3 nm

FIGURE 10.59   Illustration of the structure of GFP including the position of the chromophore in the protein structure. (The author would like to thank Richard Day for this figure from R.N. Day and M.W. Davidson, 2009. *Chem. Soc. Rev.* 38: 2887.)

molecular biologists began to prepare *in vivo* chimeric proteins containing GFP (i.e., molecular biology techniques were used to link the DNA encoding for the protein of interest to the DNA encoding for GFP), which enabled observation of the chimeric proteins using fluorescent microscopy. More recently, other classes of intrinsically fluorescent proteins have been discovered in coral such as *Discosoma*, and in other organisms. A well-characterized example is the red fluorescent protein drFP593, more commonly known as DsRed. These proteins, like the GFP family, are able to generate a fluorescent moiety out of amino acids in the primary sequence. As the name suggests, the emission from these proteins is at longer wavelengths (red-shifted) relative to GFP. It was soon found that alteration of the primary sequence of wildtype GFP could result in GFP variants with diverse spectral properties such as red- or blue-shifted emissions, higher quantum yields, and greatly improved photostability. These GFP mutants, sometimes denoted according to their colors such as YFP (yellow fluorescent protein), CFP (cyan fluorescent protein), or EGFP (enhanced green fluorescent protein), enable the tracking of multiple chimeric proteins *in vivo* and can also serve as donors and acceptors in FRET measurements. Mutations in DsRED result in similar wavelength shifts. The locations of many of the mutations made in FPs derived from jellyfish as well as corals are beautifully illustrated in **Figure 10.60** from a review by Nathan Shaner, George Patterson, and Michael Davidson (Shaner et al. (2007) *J. Cell Sci.* 120: 4247). Roger Tsien's lab also made a series of FPs named after fruit, for example, mCherry, mPlum, and so on, which have become extremely popular; these constructs are also discussed in the aforementioned review. New FPs are being reported almost every week. An article by Jones et al.

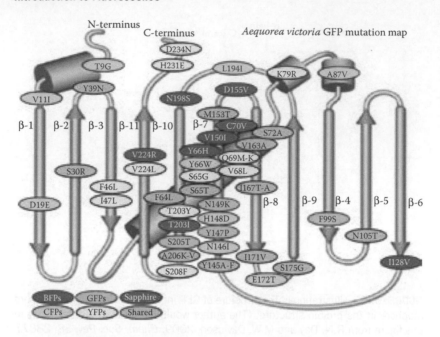

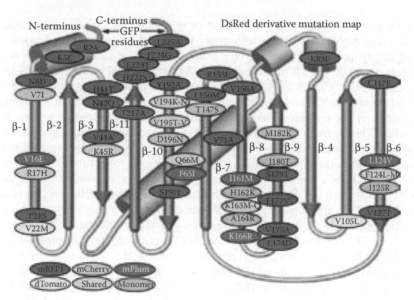

FIGURE 10.60 Illustration of the locations of many of the mutations made in FPs derived from jellyfish as well as corals. (Reproduced with permission from the review by Nathan Shaner, George Patterson and Michael Davidson (N. Shaner et al., 2007. *J. Cell Sci.* 120: 4247.))

(2012) (see Additional Reading) has a title which neatly summarizes the field, namely "A never ending race for new and improved fluorescent proteins." The field now has turned more to construction of specialized FPs that respond to specific aspects of the cellular milieu, for example, pH and redox potential. A useful review of these developments was written by S. James Remington (2011, *Protein Sci.* 20: 1509). A comprehensive, up-to-date list of FPs is very difficult to compile, since so many are developed each year. Interested readers are encouraged to consult the current literature—on a regular basis!

Another genetic method finding application in cells as well as *in vitro* is the use of the so-called FlAsH or ReAsH tags (developed by Roger Tsien and his colleagues), wherein a tetracysteine motif (such as CCPGCC, or the newer motifs HRWCCPGCCKTF and FLNCCPGCCMEP) is attached to the protein of interest, using standard genetic techniques. Then, a profluorescent compound is microinjected into the cell and attaches preferentially to the genetically introduced motif, becoming fluorescent upon the attachment. As shown in **Figure 10.61**, the profluorescent FlAsH and ReAsH are both biarsenical compounds, derivatives of fluorescein and resorufin, respectively. Upon entering the cell, these compounds will react with appropriate tetracysteine

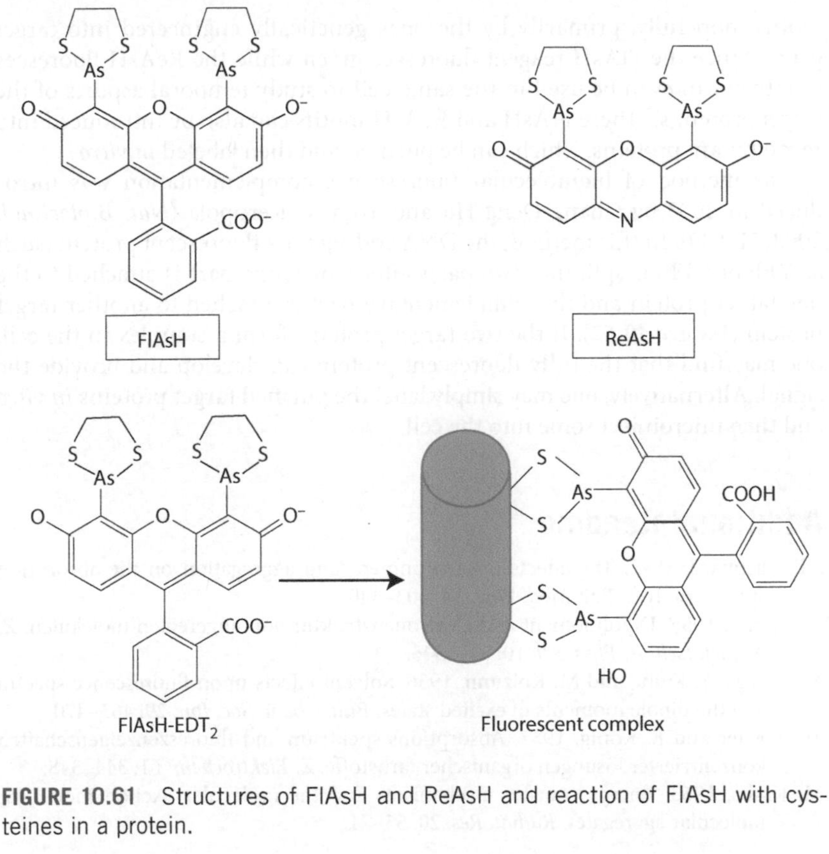

FIGURE 10.61 Structures of FlAsH and ReAsH and reaction of FlAsH with cysteines in a protein.

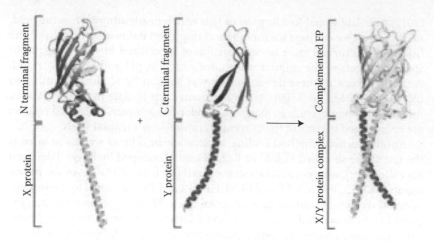

FIGURE 10.62 Illustration of the biomolecular fluorescence complementation method. (Reproduced with permission from Y. Kodama and C.D. Hu, 2012. *Biotechniques* 53: 285.)

motifs, hopefully, primarily by the ones genetically engineered into target genes. Since the FlAsH reagent fluoresces green while the ReAsH fluoresces red, they can even be used in the same cell to study temporal aspects of the target protein(s). These FlAsH and ReAsH motifs can also be introduced into recombinant proteins, which can be purified and then labeled *in vitro*.

The method of biomolecular fluorescence complementation was introduced in 2003 by Chang-Deng Hu and Tom K. Kerppola (*Nat. Biotechnol.* 2003; 21: 539). In this method, the DNA coding for a fluorescent protein (such as YFP or CFP) is split into two parts after which one part is attached to the one target protein and the complementary part is attached to another target protein (**Figure 10.62**). If the two target proteins form a complex in the cell, one may find that the fully fluorescent protein can develop and provide the signal. Alternatively, one may simply label the purified target proteins *in vitro* and then microinject some into the cell.

Additional Reading

S.E. Sheppard, 1942. The effects of environment and aggregation on the absorption spectra of dyes. *Rev. Mod. Phys.* 14: 303–340.

E. Lippert, 1955. Dipolmoment and elektronenstruktur von angeregten molekülen. *Z. Naturforsch. A. Phys. Sci.* 10: 541–545.

N. Mataga, Y. Kaifu, and M. Koizumi, 1956. Solvent effects upon fluorescence spectra and the dipole moments of excited states. *Bull. Chem. Soc. Jpn.* 29: 465–470.

Th. Förster and E. König, 1957. Absorptions spektrum and fluoreszenzeigenschaften konzentrierter lösungen organischer farbstoffe. *Z. Elektrochem.* 61: 344–348.

M. Kasha, 1963. Energy transfer mechanisms and the molecular exciton model for molecular aggregates. *Radiat. Res.* 20: 55–71.

J.E. Selwyn and J.I. Steinfeld, 1972. Aggregation equilibria of xanthene dyes. *J. Phys. Chem.* 76: 762–774.

R.W. Chambers, T. Kajiwara, and D.R. Kearns, 1974. Effect of dimer formation of the electronic absorption and emission spectra of ionic dyes. Rhodamines and other common dyes. *J. Phys. Chem.* 78: 387.

R.B. Macgregor and G. Weber, 1981. Fluorophores in polar media: Spectral effects of the Langevin distribution of electrostatic interactions. *Ann. N.Y. Acad. Sci.* 366: 140–150.

I.L. Arbeloa and P.R. Ojeda, 1982. Dimeric states of rhodamine B. *Chem. Phys. Lett.* 87: 556–560.

E. Bismuto, D.M. Jameson, and E. Gratton, 1987. Dipolar relaxations in glycerol: A dynamic fluorescence study of 2-dimethylamino-6-naphthoyl-4' cyclohexane carboxylic acid (DANCA). *J. Am. Chem. Soc.* 109: 2354–2357.

C. Reichardt, 1994. Solvatochromic dyes as solvent polarity indicators. *Chem. Rev.* 94: 2319–2358.

J.R. Lakowicz (ed.) 1994. *Topics in Fluorescence Spectroscopy* Vol. 4: *Probe Design and Chemical Sensing.*

B.D. Hamman, A.V. Oleinikov, G.G. Jokhadze, D.E. Bochkariov, R.R. Traut, and D.M. Jameson, 1996. Tetramethylrhodamine dimer formation as a spectroscopic probe of the conformation of *Escherichia coli* ribosomal protein L7/L12 dimers. *J. Biol. Chem.* 271: 7568–7573.

G. T. Hermanson, 1996. *Bioconjugate Techniques.* Academic Press, San Diego, CA.

D.M. Jameson and J.F. Eccleston, 1997. Fluorescent nucleotide analogs: Synthesis and applications. *Methods Enzymol.* 278: 363–390.

M.E. Hawkins, 2001. Fluorescent pteridine nucleoside analogs—A window on DNA interactions. *Cell Biochem. Biophys.* 34: 257–281.

A. Kawski, 2002. On the estimation of excited state dipole moments from solvatochromic shifts of absorption and fluorescence spectra. *Z. Naturforsch.* 57a: 255–262.

C.R. Cremo, 2003. Fluorescent nucleotides: Synthesis and characterization. *Methods Enzymol.* 360: 128–177.

A. Hawe, M. Sutter, and W. Jiskoot, 2007. Extrinsic fluorescent dyes as tools for protein characterization. *Pharmaceut. Res.* 25: 1487–1499.

L.D. Lavis and R.T. Raines, 2008. Bright ideas for chemical biology. *ACS Chem. Biol.* 3: 142–155.

M.A. Gilmore, D. Williams, Y. Okawa, B. Holguin, N.G. James, J.A. Ross, R.K. Aoki, D.M. Jameson, and L.E. Steward, 2011. Depolarization after resonance energy transfer (DARET): A sensitive fluorescence-based assay for botulinum neurotoxin protease activity. *Anal. Biochem.* 413: 36–42.

A.M. Jones, D.W. Ehrhardt, and W.D. Frommer, 2012. A never ending race for new and improved fluorescent proteins. *BMC Biol.*, 10: 39–41.

S.S. Wong and D.M. Jameson, 2012. *Chemistry of Protein and Nucleic Acid Cross-Linking and Conjugation.* Taylor & Francis, New York, NY.

S. Haldar and A. Chattopadhyay, 2013. Application of NBD-labeled lipids in membrane and cell biology. *Springer Ser. Fluoresc.* 13: 37–50.

C.D. Acker and L.M. Loew, 2013. Characterization of voltage-sensitive dyes in living cells using two-photon excitation. In *Chemical Neurobiology: Methods and Protocols, Methods in Molecular Biology*, Vol. 395. Ed. M. R. Banghart.

Intrinsic Protein Fluorescence

As mentioned in the previous chapter, the only naturally fluorescent amino acids are the aromatics, that is, tryptophan, tyrosine, and phenylalanine (**Figure 11.1**). For most purposes, phenylalanine can be dismissed as a useful probe, due to its low extinction coefficient and absorption maximum (~195 M⁻¹ cm⁻¹ at 258 nm) and the fact that its emission maximum is near 280 nm. Perhaps more important, though, is the fact that very few proteins have only phenylalanine residues, that is, lack tyrosine or tryptophan residues (an example of such a protein is the eubacterial ribosomal protein L7/L12, which has two phenylalanine residues). In general, as will be discussed in more detail soon, tryptophan fluorescence dominates tyrosine fluorescence, and both dominate phenylalanine fluorescence.

In 1957, Weber and his postdoctoral student, F. W. John Teale, published the first emission spectra of the aromatic amino acids and the first accurate excitation spectra. Figure 7 from their seminal paper (Teale and Weber 1957 *Biochem. J.* 65: 476) has been reproduced many times and is shown again here (**Figure 11.2**). Their published excitation spectra are reproduced in **Figure 11.3**. Anyone who has ever tried to obtain corrected excitation spectra for compounds absorbing in the UV can only admire the skill, ingenuity, and patience required to achieve their results. In the late 1950's and early 1960's, Weber and Teale published a series of important papers and communications on intrinsic protein fluorescence and the determination of absolute quantum yields. Interestingly, the quantum yield Weber and Teale reported for tryptophan, 0.20, was later found to be somewhat higher than the currently accepted value near 0.13. At the time when Weber and Teale carried out their experiments, the large temperature effect on tryptophan's lifetime and quantum yield was not appreciated. Their work, reported as done at "room temperature," was, in fact, carried out in England in the winter in a Quonset hut without central heating, which caused a marked increase in their tryptophan quantum yield relative to that expected for 25°C. Early work on intrinsic protein fluorescence was also carried out in the former Soviet Union by researchers including S.V. Konev and E.A. Burstein.

Phenylalanine Tyrosine Tyrptophan

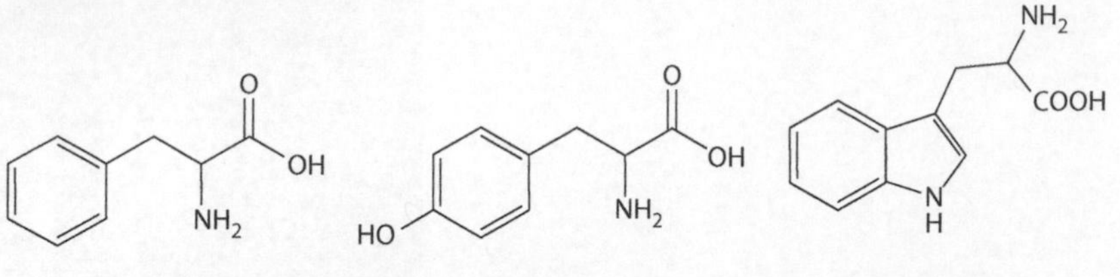

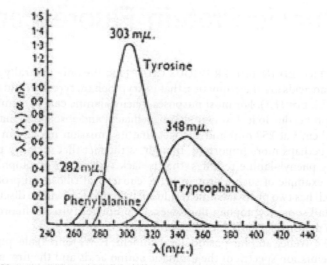

FIGURE 11.1 Structures of the aromatic amino acids.

Fig. 7. Fluorescence spectra of the aromatic amino acids in water. Abscissa: wavelength (mμ.). Ordinate: relative number of quanta.

FIGURE 11.2 Figure 7 from Weber and Teale (1957) (Reproduced with permission, from G. Weber and F.W.J.Teale, 1957, *Biochemical J.* 65, 476. © the Biochemical Society.)

Spectra

It was evident from these early studies that the intrinsic fluorescence of proteins possessing just a few, or even one, tryptophan residue and numerous tyrosine residues, were usually overwhelmingly due to the tryptophan, even in cases wherein the tyrosine residues were responsible for most of the absorption. Tryptophan usually dominates protein emission, not just because it has a higher extinction coefficient than tyrosine (5579 M^{-1} cm^{-1} at 279 nm for tryptophan versus 1405 M^{-1} cm^{-1} at 275 nm for tyrosine), but also because tyrosine can be so readily quenched in the protein matrix. In fact, some of this quenching occurs via energy transfer to tryptophan residues. An interesting case in

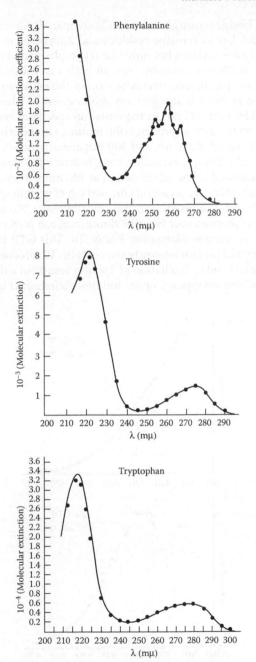

FIGURE 11.3 Excitation spectra of the aromatic amino acids from Weber and Teale (1957). (Reproduced with permission, from G. Weber and F.W.J. Teale, 1957, *Biochemical J.* 65, 476. © the Biochemical Society.)

point is that of bovine serum albumin (BSA) compared to human serum albumin (HSA). BSA has 18 tyrosine residues and 2 tryptophan residues, while HSA has 17 tyrosine residues but only one tryptophan residue. If one excites these proteins at 280 nm, one observes, in both cases, that the tryptophan emission is dominant. In fact, normalization of the emission spectra of BSA upon excitation at 280 nm and 300 nm shows essential identity, that is, no hint of a shoulder near 305 nm in the emission spectrum excited at 280 nm and hence, no evidence for a tyrosine contribution. On the other hand, a comparison of HSA excited at 280 nm and 300 nm does, in fact, exhibit a shoulder near 305 nm in the 280 nm excitation case, indicating a small tyrosine contribution (as discussed below, illumination at 300 nm will excite tryptophans but not tyrosines, therefore, a careful comparison of protein spectra excited at 280 nm and 300 nm can reveal tyrosine contributions). An exception to the dominance of tryptophan over tyrosine fluorescence in proteins occurs in the case of the *E. coli* protein Elongation Factor Tu. This GTP binding protein, involved in bacterial protein biosynthesis, contains 10 tyrosine residues and a single tryptophan residue. Excitation at 280 nm results in a distinct tyrosine contribution, which disappears upon 300 nm excitation (**Figure 11.4**). The

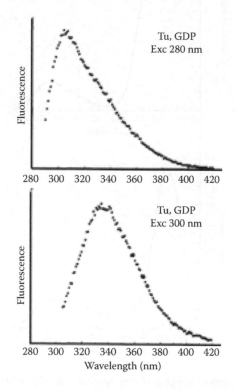

FIGURE 11.4 Emission spectra of EFTuGDP excited at 280 nm (top) and 300 nm (bottom). (Adapted with permission from D.M. Jameson et al., 1987. *Biochemistry* 26: 3894. Copyright 1987 American Chemical Society.)

most likely reason for the dominance of tyrosine emission in this single tryptophan protein is that one or more of the tyrosine residues exhibits a higher than normal quantum yield (since the lifetime, and presumably the yield, of the tryptophan residue is not unusually low).

Since these early studies, a great many papers concerned with intrinsic protein fluorescence have been published. In the 1960's, much of this work was in the nature of initial characterizations. Virtually all readily available proteins containing tryptophan and/or tyrosines were studied, and their emission properties were reported. Protein fluorescence aficionados soon realized that it was difficult dealing with the nuances of protein fluorescence, in response to ligand binding and alterations in pH and/or solvent environment, unless one could point to a single emitting residue. Hence, in the 1970s and 1980s, especially as time-resolved instrumentation became more prevalent, the hunt was on for proteins containing only a single-tryptophan. In the 1970s, site-directed mutagenesis was developed and within a decade or so it was used to remove and/or introduce tryptophan residues, with the motivation of creating single-tryptophan proteins (a strategy still popular today). This molecular biology approach, coupled with the increasing availability of commercial instrumentation, engendered a renaissance of sorts in the interest in intrinsic protein fluorescence. Unfortunately, as pointed out in 1985, in a seminal review by Joseph Beechem and Ludwig (Lenny) Brand (see Additional Reading), virtually all single tryptophan proteins examined up to that time exhibited complex excited state kinetics, that is, they could not be fit to a monoexponential decay. The reason underlying these complex decays has been debated for decades and will be addressed later.

The absorption and normalized emission spectra for the aromatic amino acids are shown in **Figure 11.5a** and **b**. The long wavelength absorption of UV light by the aromatic amino acids is due to two major $\pi-\pi^*$ absorption bands, ^1La and ^1Lb (using the nomenclature, originally proposed by John R. Platt in 1949). The absorption maxima of the lower energy bands occur near 279 nm, 275 nm, and 258 nm for tryptophan, tyrosine, and phenylalanine, respectively. The fluorescing state for tryptophan in proteins is the solvent sensitive ^1La state. In tyrosine, the ^1Lb band has the lower energy, while the higher energy ^1La band absorbs near 223 nm. The orientations of the absorption dipoles corresponding to the ^1La and ^1Lb transitions, on the aromatic frameworks, are extremely important for polarization/anisotropy measurements and will be discussed in more detail later in this chapter.

It is quite common in the literature to see correlations drawn between the position of the maximum emission wavelength (λ_{max}) of tryptophan in a protein and its exposure to the solvent. Specifically, if the λ_{max} of the emission is less than ~330 nm, the tryptophan residue is often considered to be "buried" in the protein interior. If the λ_{max} of the emission is greater than ~330 nm, then the tryptophan residue is often considered to be at least partially exposed to the solvent. Needless to say, the situation is rarely so straightforward, and such simple classifications must be taken with the proverbial grain of salt. I prefer to say that emissions near 330 nm suggest that the tryptophan residue is in an apolar or "non-relaxing" environment, which is meant to suggest that longer

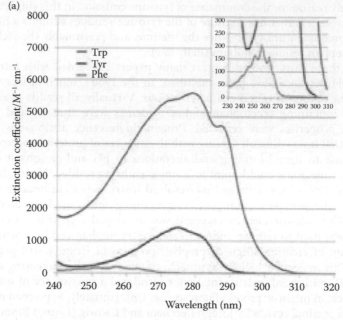

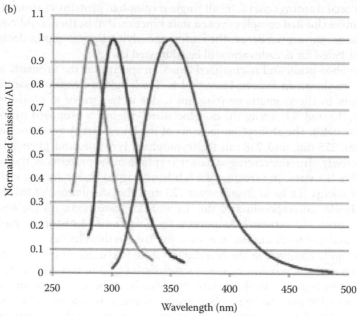

FIGURE 11.5 Absorption (a) and emission (b) spectra of the aromatic amino acids. (Reproduced from J.A. Ross and D.M. Jameson, 2008. *Photochem. Photobiol. Sci.* 7: 1301 with permission from the European Society for Photobiology, the European Photochemistry Association, and The Royal Society of Chemistry.)

emissions may not be due to reorientation of water dipoles (see discussion on dipolar relaxation in Chapter 10), but could perhaps be due to relaxation of the protein matrix around the excited state dipole of tryptophan. In fact, even this consideration is simplistic and now we better understand the complexities of the photophysics of tryptophans in proteins. The microscopic details of the environment such as the extent of polarizability, the presence of water molecules, and the specific interactions between the indole ring and polar groups in the protein will all influence the maximum emission wavelength. As Gregorio Weber used to point out to me, the environments of residues in a protein matrix are highly anisotropic and, as such, can never really be approximated by isotropic solvents.

In a few cases, proteins without tryptophan have been shown to emit near 340 nm. These rare cases are due to tyrosinate emission, that is, from tyrosine with an ionized phenolic residue. The pKa of the phenolic group in the ground state is about 10.3, yet, it is about 4.2 in the first singlet excited state. Hence, if the tyrosine residue is in an environment that can facilitate extraction of the phenolic proton during the excited state, it can emit as the tyrosinate moiety. If tyrosinate fluorescence is suspected, one can denature the protein with urea or GuHCl and see if the peak near 340 nm blue-shifts back near 305 nm.

Polarization

Many polarization/anisotropy studies on intrinsic protein fluorescence have been carried out over the past 50 years. The goals of these studies were diverse, but interpretation of almost any result required knowledge of the underlying polarization properties of the indole moiety. Weber first published the excitation polarization spectrum of indole in 1960 and then, with Bernard Valeur in 1977 (see Additional Reading), published a higher resolution spectrum, which is the cornerstone of protein polarization studies (Figure 11.6). The orientations of the ^{1}La and ^{1}Lb transitions on the indole structure are depicted in Figure 11.7. A glance at these data indicates that care must be taken when choosing the excitation wavelength—and slitwidths—for polarization/anisotropy studies on proteins. It is a common practice, for example, to excite a protein at 295 nm in order to excite primarily tryptophan, but not tyrosine, residues. Examination of the inset in Figure 11.5a indicates that this approach may work if there are only a few tyrosines, although the excitation slitwidth should be as narrow as practical. If one plans on carrying out polarization studies, though, one sees from Figure 11.5 that a small wavelength shift, up or down from 295 nm, is enough to cause a significant change in the limiting polarization. For this reason, it is better to excite the protein at 300 nm, on the polarization plateau, so that one can be more certain about the limiting polarization value. In general, 300 nm excitation of proteins is also preferable if one wants to take tyrosine out of consideration—especially considering the possibility of tyrosine to tryptophan energy transfer (see below). On a practical note, since the absorption of tryptophan is fairly low at 300 nm (Figure 11.5), one should be careful with contributions from the buffer background including the Raman peak, which at 300 nm excitation will be at 334 nm.

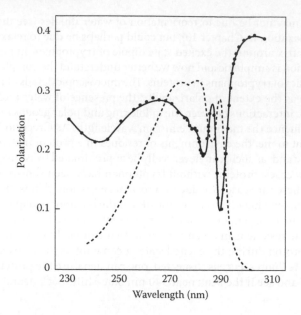

FIGURE 11.6 Excitation polarization spectrum (solid line) and excitation spectrum (broken line) of indole in propylene glycol at −58C. (From D.M. Jameson et al., 1978. *Rev. Sci. Instru.* 49: 510.)

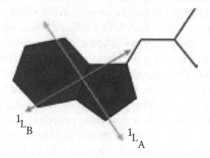

FIGURE 11.7 Orientation of the ^1L_A and ^1L_B transition dipoles on the ring system of indole. (The author would like to thank Jay R. Knutson for this figure.)

Lifetime

In 1961, based on considerations of oscillator strength, quantum yields, and polarization measurements, Weber speculated that the lifetime of tryptophan in solution would be about 2.5–4 ns. Instrumentation capable of measuring tryptophan lifetimes did not appear until the mid-1960s, and when such measurements were carried out, it was clear that Weber's estimations were correct. The fluorescence lifetime and the number of lifetime components of tryptophan in aqueous solution are strongly dependent on the pH of the solution. **Figure 11.8** shows the variation of the phase lifetime value (at 10 MHz modulation frequency) of tryptophan as a function of pH (I note that studies on the variation of emission intensity of tyrosine and tryptophan as a function of pH were first published in 1958 by Audrey White). At neutral pH and 20°C, tryptophan exhibits two lifetime components, one major component of ~3 ns with an emission maximum at 350 nm, and one minor component of ~0.5 ns with an emission maximum near 335 nm. This critical observation was first made by Arthur Szabo and David Rayner in 1980 (see Additional Reading). They postulated that this biexponential decay of tryptophan at neutral pH is due to three possible "rotamers" of the indole rings about the Cα–Cβ bond within tryptophan. These three rotamers have been labeled I, II, and III and their Newman projection formulae are shown in **Figure 11.9**. Rotamer I is the preferred orientation due to the proximity of the positive charge on the α-ammonium group. Rotamer II is the least energetically favorable as a result of the electrostatic repulsion between the deprotonated carboxylic acid group and the π-electron cloud of the indole rings. While Rotamer III is more energetically favorable than Rotamer II, it may suffer from steric hindrance from the close proximity of the Cα. This model has been labeled the *classical rotamer model*. At higher pH levels, when the amino group becomes protonated, the 3 ns component is replaced by a longer component near 8.7 ns. As the pH is raised above the pKa of the amino group, the relative contribution

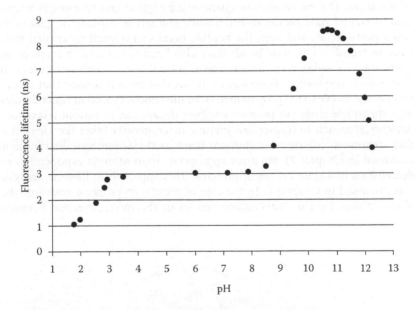

FIGURE 11.8 Plot of the fluorescence lifetime of tryptophan as a function of pH. Data points represent phase lifetime at 10 MHz light modulation frequency. (From PhD thesis of D.M. Jameson.)

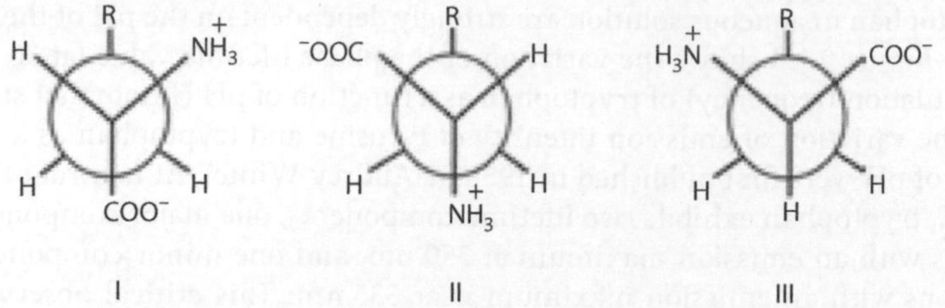

FIGURE 11.9 Newman projection formulas, looking along the C_α–C_β bond, depicting the three rotational rotamers of tryptophan. R represents the indole ring.

of the 3 ns or 8.7 ns component follows the proportion of the protonated or deprotonated form.

The emission maxima of tryptophan in proteins ranges from 308 nm (for the copper containing protein azurin) to 352 nm. The excited state lifetimes range from several picoseconds to nearly 10 ns and the quantum yields vary from near zero to around 0.35. N-acetyl-L-tryptophanamide (NATA) (**Figure 11.10**) is typically used in solution as an analog of a tryptophan residue within a protein. NATA emission decays with a single exponential of 3.1 ns at 20°C. The lifetime of tryptophan in proteins is affected by many processes, which enhance the nonradiative decay rate, including solvent quenching, excited state proton transfer, excited state electron transfer, intersystem crossing, and temperature. The mechanism of quenching by lysine and tyrosine is attributed to excited state proton transfer, while glutamine, asparagine, glutamic, and aspartic acids, and even the peptide bond can quench by excited state electron transfer. Disulfide bonds have also been implicated in tryptophan quenching. In addition to these residues, solvent (e.g., water) molecules can also quench tryptophan fluorescence. It has also been proposed that movement of the dipole in the protein matrix, or movement of solvent water dipoles near the indole ring, can produce complex decays due to solvent relaxation. Another approach to tryptophan lifetime heterogeneity takes the viewpoint that continuous lifetime distributions (such as the Lorentzian distributions discussed in Chapter 7), are more appropriate than discrete exponentials to describe excited state decays in proteins. This approach to lifetime analysis was discussed in Chapter 6. In the case of protein tryptophan emission, the physical basis for the distributions resides in the interconversion between

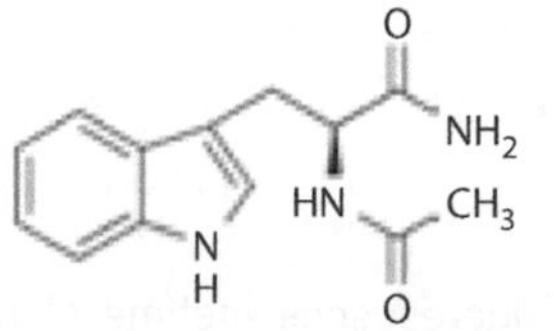

FIGURE 11.10 Structure of N-acetyl-L-tryptophanamide (NATA).

conformations, each characterized by a quasi-continuum of energy substates, which place the tryptophan residue in different environments. The observed lifetime heterogeneity, according to this line of reasoning, is thus a function of the interconversion rates and hence assignment of a discrete lifetime component to a particular protein conformation may be an oversimplification (see Alcala et al. 1987 in Additional Reading). As mentioned already, HSA contains a single tryptophan residue (Trp 214), and, not surprisingly, has been the subject of numerous fluorescence studies, including time-resolved studies. Figure 6.8 showed the Lorentzian distribution used to fit the excited state lifetime data for HSA. I should note that such studies on HSA require care that the monomeric protein is isolated from the covalent aggregates that normally form in commercially available lyophilized HSA—failure to purify the monomeric HSA from the higher aggregates formed by sulfhydryl exchange will lead to erroneous conclusions, since the dimer has different lifetime properties than the monomer.

Since this treatise is an introduction, I shall resist the temptation to give a more detailed exposition on theories concerning the influences on the excited state lifetime of tryptophan in proteins. Intrepid readers should consult the copious primary literature on that topic. Several relevant articles on this topic are listed under Additional Reading at the end of this chapter.

Electronic Energy Transfer in Proteins

As early as 1960, Gregorio Weber showed that electronic energy transfer— or in today's parlance, FRET—could take place from tyrosine to tryptophan (heterotransfer) as well as between tryptophan residues (homotransfer). The demonstration of these FRET events relied on polarization measurements. As the reader will no doubt recall from Chapter 8, FRET between two dipoles of differing orientations will lead to depolarization of the acceptor emission, compared to direct excitation of the acceptor. Hence, a comparison of the polarization of the tryptophan emission, excited at 270 nm (which will excite tyrosine as well as tryptophan residues) and excited at 305 nm (which will excite only tryptophan) can provide evidence of FRET. As Weber showed, in the absence of tyrosine to tryptophan FRET, the ratio of the polarizations observed upon 305 nm excitation and 270 nm excitation, that is, the 305/270 ratio, will be in the range of 1.4–1.5. Tyrosine to tryptophan energy transfer (upon 270 nm excitation), however, leads to an increase in this ratio due to the decrease in the tryptophan polarization after FRET from tyrosine. In such cases, 305/270 ratios as high as 3 have been observed.

Tryptophan to tryptophan energy transfer can also be detected using polarization measurements. The Förster critical transfer distance for tryptophan to tryptophan transfer is around 6–12 Å. To detect such homoFRET, one can compare the polarization observed upon exciting the protein at 295 nm and at 310 nm. At 295 nm, tyrosine excitation is minimized, but at this excitation wavelength, tryptophan to tryptophan FRET can occur. Upon excitation at 310 nm, however, the extent of homotransfer is greatly reduced, due to the Weber Red-Edge effect (mentioned in Chapter 8). Hence, the 295 nm excitation essentially carries information on the depolarization of the tryptophan due to rotation and homoFRET, while the 310 nm excitation carries information on the depolarization due to rotation alone. For tryptophan in glycerol at low temperatures (where there is no rotation), the 310/295 ratio of the polarizations is near 1.7, whereas in cases such as ditryptophan and polytryptophan, this ratio is greater than 2 (Moens et al. 2004 *The Protein J.* 23: 79). Similar increases in this 310/295 ratio have been observed in protein systems, although this approach is not well-known. An example of the application of this method will be given in the next section.

Another interesting case of FRET with intrinsic protein fluorescence takes place with proteins containing heme groups, such as hemoglobin, myoglobin, horseradish peroxidase, and the cytochromes. Hemoglobin, as the reader will recall, is a tetramer composed of two different types of subunits, termed α and β, each of which is about 16 kDa. The α subunit has one tryptophan residue while the β subunit has two. The heme moiety has a significant absorption where tryptophan emits, and hence, given the fact that there are four heme groups in the tetramer, the extensive energy transfer is not unexpected. Removal of the heme group leads, of course, to a large increase in the intrinsic tryptophan fluorescence. Initial studies of the emission spectra and quantum yields of hemoglobin were carried out in Weber's lab and indicated quantum yields less than 1% relative to tryptophan. These types of studies are very difficult to realize since the absorption of the heme group makes inner filter effects a concern as well as buffer background and Raman peaks. The intrinsic fluorescence of hemoglobin has also been extensively studied by Rhoda Hirsch (e.g., R.E. Hirsch and R.L. Nagel 1981 *J. Biol. Chem.* 256: 1080), using front-face methodologies, which mitigate inner filter effects and, hence, allows the use of higher heme protein concentrations than with traditional right angle optics (as discussed in Chapter 3). The emission of myoglobin, a monomer, is also highly quenched. Horseradish peroxidase (HRP), a monomeric glycoprotein of about 44 kDa, which contains one heme group and one tryptophan residue, exhibits a much higher intrinsic fluorescence than hemoglobin or myoglobin. The tryptophan fluorescence of HRP is quenched only about 85% relative to apohorseradish peroxidase. The situation with cytochromes varies depending on the protein. In ferricytochrome c, for example, at neutral pH, the single tryptophan residue (W59) has a quantum yield of about 2% relative to free tryptophan, while in the presence of 9 M urea, the yield of this tryptophan residue increases to about 65% relative to free tryptophan (Tsong 1974 *J. Biol. Chem.* 249: 1988). Evidently, unfolding of the cytochrome c leads to greatly reduced FRET between the tryptophan residue and the heme group (which is covalently bound to the protein matrix). Transferrin, an iron-binding protein, also demonstrates quenching of the tryptophan emission, even though there is no heme moiety present. In this protein, a single-chain polypeptide of 629 amino acids, two irons are bound by specific tyrosine, histidine, and aspartic acid residues. When iron is bound, a ligand to metal charge-transfer (LMCT) band appears, when the π-orbitals of the liganding tyrosine residues merge with the d-orbital of the iron atom, creating a new orbital delocalized between the tyrosine and iron. This LMCT band absorbs in the visible as well as the UV, and results in quenching of the intrinsic protein fluorescence via FRET (James et al. 2010 *Protein Sci.* 19: 99).

Use of Site-Directed Mutagenesis

Site-directed mutagenesis has been widely used to alter the number of tryptophan residues in proteins or to alter the amino acid residues near tryptophan residues. Tryptophan residues have been inserted and also removed from proteins, usually with the goal of constructing single-tryptophan systems with the tryptophans placed in regions of interest. An example of such a study is given in **Figure 11.11**, which shows the intrinsic emission for a kinase containing four tryptophan residues, along with the spectra of the four single tryptophan constructs (from Watanabe et al. 1996 *Prot. Sci.* 5: 904). The perceptive reader may note the Wood's Anomalies near 370 nm in these spectra, which were obtained using an SLM monochromator. One should also note that the intensities of the four single tryptophan residues sum to greater than the four-tryptophan wildtype protein. The 310/295 ratios for several single and double tryptophan constructs of this kinase (from Helms et al. 1998 *Biochemistry* 37: 14057) are shown in **Figure 11.12**. As is evident, residues W299 and W320 demonstrate FRET, which indicates their proximity in the kinase three-dimensional structure. I must add that these types of experiments are rather difficult and one must be *extremely* careful to eliminate stray light, for example, Rayleigh Ghosts, from the measurement. This precaution usually requires

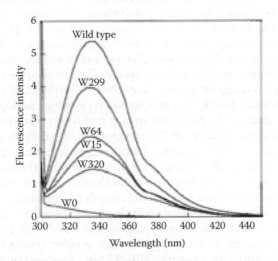

FIGURE 11.11 Emission spectra of wildtype and single tryptophan mutants of fructose 6-phosphate, 2-kinase:fructose 2,6-bisphosphatase. (Modified from F. Watanabe et al., 1996. *Protein Sci.* 5: 904.)

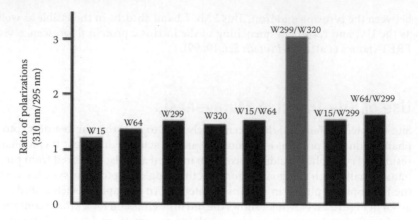

FIGURE 11.12 Ratios of polarizations for the tryptophan emissions for single and double tryptophan proteins excited at 295 nm and 310 nm. (Adapted with permission from M.K. Helms et al., 1998. *Biochemistry* 37: 14057. Copyright 1998 American Chemical Society.)

either double monochromators or the use of interference filters in conjunction with a single monochromator.

Tryptophan Analogs

Incorporation of tryptophan analogs into proteins for elucidation of metabolic pathways, including protein biosynthesis, was carried out as early as the 1950s. The first spectroscopic characterization of a protein containing tryptophan analogs was in 1968 by Sondra Schlesinger who introduced 7-azatryptophan into alkaline phosphatase and noted the alteration in the protein's absorption and fluorescence spectra (discussed below). In the 1990s, several groups (principally those of Jacob Petrich, Arthur Szabo, J. Alexander Ross, and Maurice Eftink) began to utilize tryptophan analogs for detailed fluorescence studies. The particular analogs that have proven to be most popular in fluorescence studies are 7-azatryptophan and 5-hydroxytryptophan, shown in **Figure 11.13**. To prepare the recombinant proteins, an *Escherichia coli* auxotroph is typically grown on medium containing the analog of choice. Appropriate expression vectors are essential to this process, and readers with a sustaining interest in this topic should consult the literature.

The great utility of these analogs is that their absorption spectra extend further to the red compared to tryptophan (**Figure 11.13**). Hence, one can excite these residues at wavelengths where regular tryptophan does not absorb significantly, for example, 310–315 nm. This selectivity allows one to study the analog fluorescence even in the presence of other tryptophan containing proteins. An additional useful feature is that the emission from

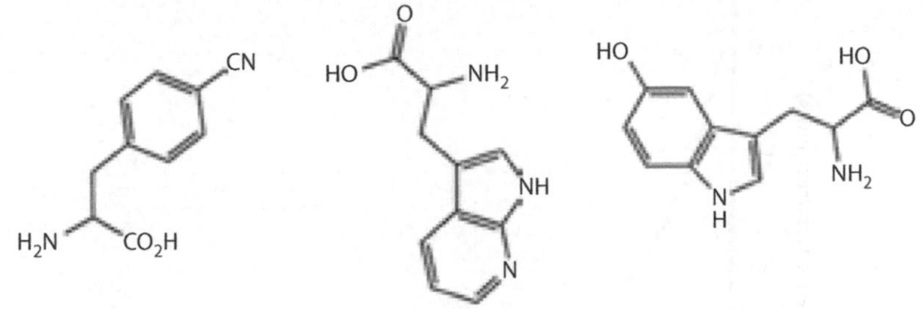

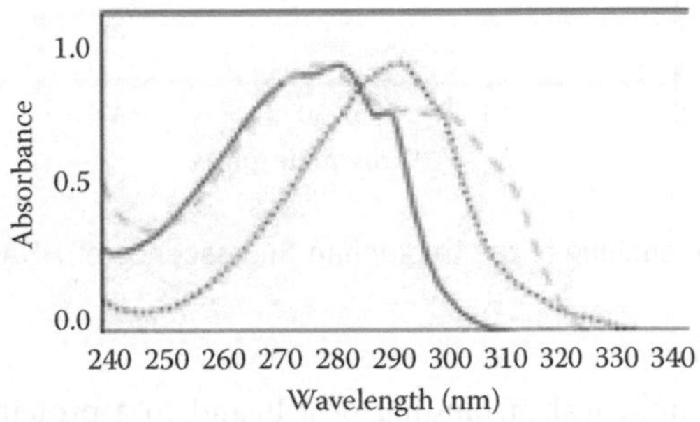

FIGURE 11.13 Structures of p-cyanophenylalanine, 7-azatryptophan, 5-hydroxyltryptophan and absorption spectra of tryptophan (blue solid line), 7-azatryptophan (red dotted line), and 5-hydroxyltryptophan (green dashed line).

7-azatryptophan is highly sensitive to its surroundings. Its quantum yield, although low in water, can increase dramatically in less polar environments, for example, from ~0.01 in water to 0.25 in acetonitrile. More recently, p-cyanophenylalanine (p-CN) (Figure 11.13) has also been introduced as an analog for phenylalanine. p-CN can be excited out to 280 nm and its emission in the 290–310 nm region allows it to be used in FRET experiments with tryptophan.

Applications

Perhaps, the most common application of intrinsic protein fluorescence is to detect and monitor conformational changes in the protein matrix. These changes may be due to ligand binding, protein oligomerization or dissociation, binding to DNA or membranes, protein unfolding or refolding—just about any molecular event which involves a protein. I shall give a few examples from the literature. I apologize for presenting so many examples from studies in which I was involved, but since I am familiar with these examples, they are easier to discuss and they serve to make my points.

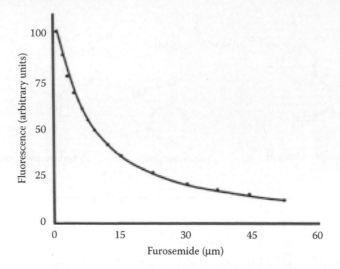

FIGURE 11.14   Quenching of the tryptophan fluorescence of HSA by furosemide.

Ligand Binding

In many cases, noncovalent binding of a ligand to a protein can result in changes in one, or most likely more than one, of the intrinsic protein fluorescence parameters. These changes may be due to either conformational changes in the protein structure, for example, which alter a tryptophan's local environment, or in some cases due to FRET, for example, from a tryptophan residue to the ligand. Of course, both of these mechanisms may also operate together. In the example shown in **Figure 11.14**, the tryptophan fluorescence of HSA (recall that HSA has a single tryptophan residue) was quenched upon binding of the drug furosemide (a diuretic). Furosemide absorbs at 330 nm, and the quenching of the HSA tryptophan is presumably due to an energy transfer process. As is evident from **Figure 11.14** (from Voelker et al. 1989 *J. Pharm. Exp. Therap.* 50: 772), one can use these types of data to readily determine the binding constant under various conditions, for example, ionic strength. Another example of using tryptophan fluorescence to follow binding is shown in **Figure 11.15** (from Daniel and Weber 1966 *Biochemistry* 5: 1893), which shows that the protein's (BSA) tryptophan fluorescence is quenched as ANS binds—the concomitant increase in the ANS fluorescence upon binding is also shown. Note that the wavelength scale runs from higher to lower wavelengths as we look left to right, which is opposite from most spectra shown today. This orientation was due to the fact that the spectra were taken using a stripchart recorder, which was rewound after each ANS addition. The keen observer will note the Rayleigh peak at 280 nm and the second-order peak near 560 nm. Finally, **Figure 11.16** shows the use of intensity and anisotropy to follow binding of a ligand (fructose-6-phosphate) to a single tryptophan mutant of fructose 6-phosphate, 2-kinase/fructose 2,6-bisphosphatase (from Helms et al. 1998 *Biochemistry*, 37: 14057).

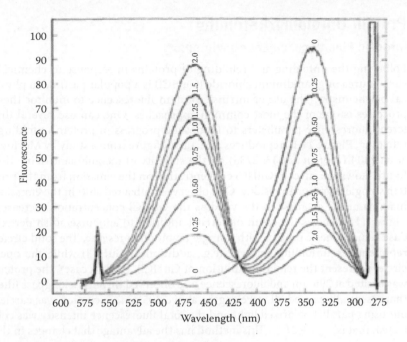

FIGURE 11.15 Quenching of the tryptophan fluorescence of BSA by ANS. (From E. Daniel and G. Weber, 1966. *Biochemistry* 5: 1893. Copyright 1966 American Chemical Society.)

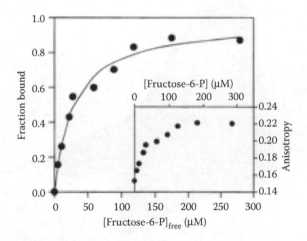

FIGURE 11.16 Effect of fructose-6-phosphate on the intensity and anisotropy of a single tryptophan mutant of fructose 6-phosphate, 2-kinase/fructose 2,6-bisphosphatase. (Adapted with permission from M.K. Helms et al., 1998. *Biochemistry* 37: 14057. Copyright 1998 American Chemical Society.)

Protein Unfolding/Refolding

Intrinsic Fluorescence Spectra/Intensity

Following the unfolding and refolding of proteins in response to chemicals such as urea or guanidinium chloride (GuHCl) is a popular pastime in physical biochemistry. The use of intrinsic protein fluorescence to monitor these processes is one of the most common approaches. One can use several different fluorescence parameters to follow the progress of protein unfolding/refolding. **Figure 11.17** reproduces a beautiful figure from a study by McHugh et al. 2004 (*Protein Sci.* 13: 2736) on the stability of recombinant ricin. This figure shows the effect of GuHCl concentration on the emission from the protein's single tryptophan residue. One observes both a red shift in the emission maxima and a decrease in the yield as the GuHCl concentration increases. **Figure 11.18** shows the result of GuHCl addition to apohorseradish peroxidase, a monomeric protein with a single tryptophan residue. The solid circles represent the unfolding pathway (e.g., addition of GuHCl) while the open circles represent the refolding (dilution of GuHCl). In both cases, the protein was excited at 295 nm and fluorescence was observed through a longpass filter that essentially passed all wavelengths above 305 nm. Excitation was carried out using parallel polarized light and the total fluorescence intensity was collected, that is, $I_{para} + 2I_{perp}$. This method has the advantage that changes in the

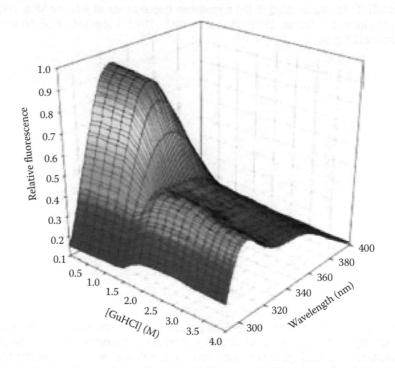

FIGURE 11.17 Effect of guanidinium chloride on the emission of ricin's single tryptophan residue. (From C.A. McHugh et al., 2004. *Protein Sci.* 13: 2736.)

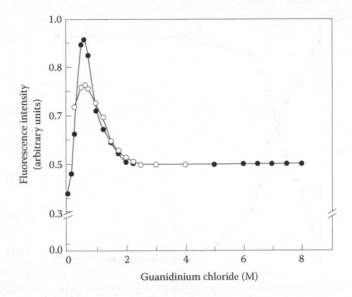

FIGURE 11.18 Effect of guanidinium chloride on the intensity of the single tryptophan emission of apohorseradish peroxidase. Solid circles and open circles indicate unfolding and refolding pathways, respectively. (From M. Lasagna et al., 1999. *Biophys. J.* 76: 443.)

polarization or emission maximum does not bias the intensity readings. In cases such as these, wherein the emission maximum shifts to longer wavelengths (red shifts) as the protein unfolds (which is exactly what happens in this case as indicated in **Figure 11.18**), care must be exercised if one follows the process at a fixed wavelength through a monochromator. In these cases, an isoemissive wavelength should be used (unless, of course, the aim is to find a condition which maximizes the signal change). An instrument bias still persists, though, if the polarization of the emission changes with unfolding/refolding since, as previously mentioned, monochromators typically respond differently to light of different polarizations.

Polarization

As shown in **Figure 11.19**, polarization also can be used to follow the unfolding process. As the apohorseradish peroxidase unfolds, the local mobility of the tryptophan residue increases, which leads to lower polarization as shown. One notes that the emission maximum also red-shifts during unfolding.

Lifetimes

Time-resolved fluorescence can also be used to follow protein unfolding/refolding as shown in **Figure 11.20** for the apohorseradish peroxidase case. In this case, the lifetime data were fit to a single Lorentzian distribution, and

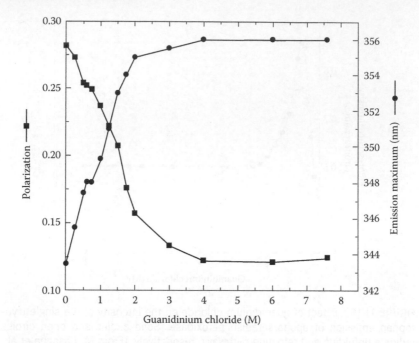

FIGURE 11.19 Effect of guanidinium chloride on the polarization and emission maximum of apohorseradish peroxidase. (From M. Lasagna et al., 1999. *Biophys. J.* 76: 443.)

both the center of this distribution and its width were plotted as a function of GuHCl concentration. As before, closed circles represent increasing GuHCl concentrations while open circles represent decreasing concentrations.

Protein–Protein and Protein–DNA Interactions

Macromolecular interactions—such as protein–protein or protein–DNA—can, in some cases, be followed using intrinsic protein fluorescence. An interesting application of intrinsic protein fluorescence is to follow the interaction of proteins with DNA. **Figure 11.21** (modified from Favicchio et al. 2009 *Methods Mol. Biol.* 543: 589) shows that interaction of the DNA binding domain of Sox-5 protein to 12 bp DNA leads to quenching of the protein's tyrosine fluorescence. Figure 6.26, in Chapter 6, showed the time-resolved studies of Elongation Factor Tu, free and bound to Elongation Factor Ts. As pointed out, in EFTuGDP, the trytophan residue had only limited local mobility, whereas upon binding to EF-Ts, the local tryptophan mobility increases markedly (note all data were collected using 300 nm excitation so that tyrosine residues were not excited).

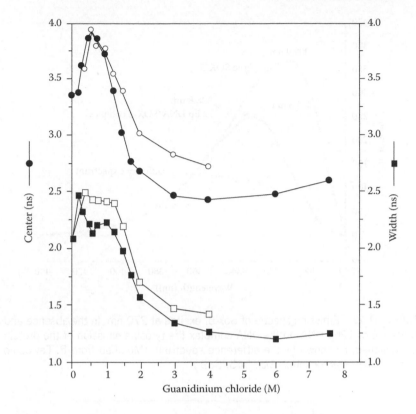

FIGURE 11.20 Effect of guanidinium chloride on the center and distribution width of the lifetime apohorseradish peroxidase. Solid circles and open circles indicate unfolding and refolding pathways, respectively. (From M. Lasagna et al., 1999. *Biophys. J.* 76: 443.)

Tyrosine Studies

Studies on tyrosine fluorescence in proteins are less common than tryptophan studies, given the dearth of proteins without tryptophan residues. Also, tyrosine is less sensitive to its environment in proteins than tryptophan. Consequently, lifetime analysis offers more possibilities in these systems. **Figure 11.22** (from Ferrira et al. 1994 *Biophys. J.* 66: 1185) shows the absorption, excitation, and emission spectra for the single tyrosine residue in bovine erythrocyte Cu, Zn superoxide dismutase (BSOD), a homodimer containing a single solvent exposed tyrosine residue per subunit. **Figure 11.23** shows the results of temperature studies on the lifetime of this residue, which was analyzed using a unimodal Lorentzian lifetime distribution model.

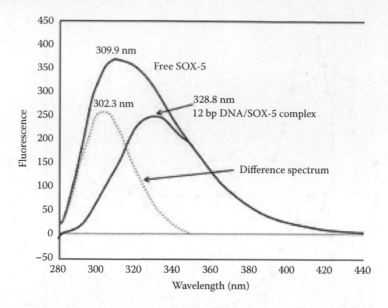

FIGURE 11.21 Emission spectra of Sox-5, excited at 270 nm, in the absence and presence of 12bp DNA. In the DNA complex the tyrosine emission of the protein is quenched, as shown by the difference spectrum. (Modified from R. Favicchio et al., 2009. *Methods Mol. Biol.* 543: 589.)

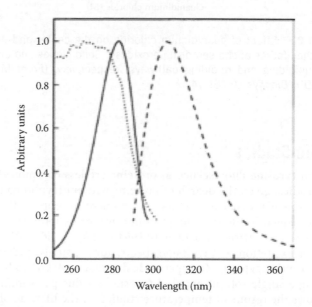

FIGURE 11.22 Absorption (dotted line), excitation (solid line) and emission (dashed line) spectra of the single tyrosine residue of bovine superoxide dismutase. (From S.T. Ferreira et al., 1994. *Biophys. J.* 66: 1185.)

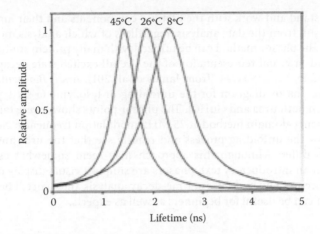

FIGURE 11.23 Fluorescence lifetime distributions of bovine superoxide dismutase at various temperatures. The curves are normalized for a relative amplitude of 1 at the center of the lifetime distribution. (From S.T. Ferreira et al., 1994. *Biophys. J.* 66: 1185.)

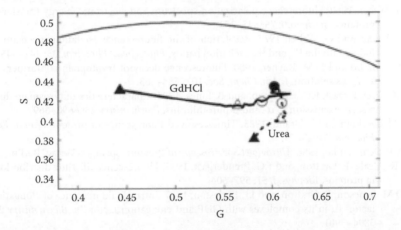

FIGURE 11.24 Phasor diagram for the unfolding of lysozyme (a six tryptophan protein) in both urea and GuHCl. Concentrations of 0 M (closed circle), 1 M (open circle), 2 M (open triangle), and 6/8 M (closed triangle) GdHCl and urea are highlighted. (From N.G. James et al., 2011. *Anal. Biochem.* 410: 70.)

Phasors and Protein Fluorescence

Finally, I would like to draw attention to the relatively new approach to time-resolved studies on intrinsic protein fluorescence, namely, the phasor method described in Chapter 6. As mentioned earlier, even single tryptophan proteins exhibit complex excited state decays. As a consequence, it is often difficult to understand and work with the lifetime components and their amplitudes, which result from the data analysis, regardless of which analysis method one is using. The phasor method can be utilized with many protein studies to simply provide a visual representation of the overall excited state decay process. For example, **Figure 11.24** (from James et al. 2011 *Anal. Biochem.* 410: 70) shows the phasor diagram for the unfolding of lysozyme (a six tryptophan protein) in both urea and GuHCl. The phasor points shown were taken using the frequency domain method at 25 MHz modulation frequency; clearly, one can follow the unfolding process and readily see that the urea and GuHCl pathways differ. Although this approach may seem somewhat esoteric to include in an introductory text, phasors are simply a visual display of the raw data that require no expertise at time-decay analysis and hence, I believe this approach can be useful for beginners as well as experts.

Additional Reading

F.W.J. Teale and G. Weber, 1957. Ultraviolet fluorescence of the aromatic amino acids. *Biochem. J.* 65: 476–482.

F.W.J. Teale, 1960. The ultraviolet fluorescence of proteins in neutral solutions. *Biochem. J.* 76: 381–388.

G. Weber, 1960. Fluorescence polarization spectrum and electronic energy transfer in proteins. *Biochem. J.* 75: 345–352.

B. Valeur and G. Weber, 1977. Resolution of the fluorescence excitation spectrum of indole into the ^1L_a and ^1L_b excitation bands. *Photochem. Photobiol.* 25: 441–445.

A.G. Szabo and D.M. Rayner, 1980. Fluorescence decay of tryptophan conformers in aqueous solution. *J. Am. Chem. Soc.* 102: 554–563.

F.G. Prendergast, P.D. Hampton, and B. Jones, 1984. Characteristics of tyrosinate fluorescence emission in α- and β-purothionins. *Biochemistry* 23: 6690–6697.

J.M Beechem and L. Brand, 1985. Time-resolved fluorescence of proteins. *Annu. Rev. Biochem.* 54: 43–71.

A.P. Demchenko, 1986. *Ultraviolet Spectroscopy of Proteins.* Springer-Verlag, Berlin.

J.R. Alcala, E. Gratton, and F.G. Prendergast, 1987. Fluorescence life time distributions in proteins. *Biophys. J.* 51: 597–604.

D.M. Jameson, E. Gratton, and J.F. Eccleston, 1987. Intrinsic fluorescence of elongation factor Tu in its complexes with GDP and elongation factor Ts. *Biochemistry* 26: 3894–3901.

S.T. Ferriera, L. Stella, and E. Gratton, 1994. Conformational dynamics of bovine Cu, Zn superoxide dismutase revealed by time-resolved fluorescence spectroscopy of the single Tyrosine residue. *Biophys. J.* 66: 1185–1196.

F. Watanabe, D.M. Jameson, and K. Uyeda, 1996. Enzymatic and fluorescence studies of four single tryptophan mutants of rat testis fructose 6-phosphate, 2-kinase:fructose 2,6-bisphosphatase. *Protein Sci.* 5: 904–913.

M.K. Helms, T.L. Hazlett, H. Mizuguchi, C.A. Hasemann, K. Uyeda, and D.M. Jameson, 1998. Site-directed mutants of rat testis fructose 6-phosphate, 2-kinase:fructose 2,6-bisphosphatase: Localization of conformational alterations induced by ligand binding. *Biochemistry* 37: 14057–14064.

Y. Chen and M.D. Barkely, 1998. Toward understanding tryptophan fluorescence in proteins. *Biochemistry* 37: 9976–9982.

A. Sillen and Y. Engelborghs, 1998. The correct use of "average" fluorescence parameters. *Photochem. Photobiol.* 67: 475–486.

M. Lasagna, E. Gratton, D.M. Jameson, and J.E. Brunet, 1999. Apohorseradish peroxidase unfolding and refolding: Intrinsic tryptophan fluorescence studies. *Biophys. J.* 76: 443–450.

S.M. Twine and A.G. Szabo, 2003. Fluorescent amino acid analogs. *Methods Enzymol.* 360: 104–127.

P.D. Moens, M.K. Helms, and D.M. Jameson, 2004. Detection of tryptophan to tryptophan energy transfer in proteins. *Protein J.* 23: 79–83.

J.A. Ross and D.M. Jameson, 2008. Frequency domain fluorometry: Applications to intrinsic protein fluorescence. *J. Photochem. Photobiol. Sci.* 7: 1301–1312.

N.G. James, J.A. Ross, A.B. Mason, and D.M. Jameson, 2010. Excited-state lifetime studies of the three tryptophan residues in the N-lobe of human serum transferrin. *Protein Sci.* 19: 99–110.

N.G. James, J.A. Ross, M. Stefl, and D.M. Jameson, 2011. Application of phasor plots to *in vitro* protein studies. *Anal. Biochem.* 410: 70–76.

C.P. Pan, P.L. Muiño, M.D. Barkley, and P.R. Callis, 2011. Correlation of tryptophan fluorescence spectral shifts and life times arising directly from heterogeneous environment. *Methods Enzymol.* 487: 1–38.

Appendix: Rogue's Gallery of Fluorescence Artifacts and Errors

IN THIS APPENDIX to the main text, I discuss the most common artifacts, errors, and misunderstandings encountered in fluorescence measurements. Many of these have been discussed earlier, but I believe it useful to present a rogue's gallery of potential heartbreakers in one place in this book. I shall list them loosely in the order that I believe reflects their relative importance, although this order is naturally somewhat subjective.

The Raman Peak

In my experience, the Raman peak is by far the most common feature to befuddle the novice. The annoyance the Raman peak causes fluorescence practitioners can only be matched by the annoyance fluorescence causes the practitioners of Raman spectroscopy (I suppose the fluorescence community can take some solace in this observation). Since the majority of readers of this book are likely concerned with fluorescence in aqueous solutions, the most relevant Raman peak, for our consideration, is that due to water. At fluorescence workshops, I often ask students if anyone knows the molarity of water— usually only the chemists among them know that it is 55 M (once a student answered 30 M and since we were in Italy, I replied, "No, you're thinking of grappa)!" Even though the water Raman emission is intrinsically very weak, given the high concentration of water, it can be significant in some circumstances. As discussed in Chapter 4, the O–H stretching frequencies occurs at about 3400 cm^{-1}, and hence, one can easily calculate the wavelength expected for a water Raman peak as a function of the excitation wavelength, as was shown in Figure 4.7.

When a Raman peak is suspected one can, of course, substitute the solvent minus fluorophores for the sample, with the same instrument settings used for the sample run. One should then readily see the Raman peak in the solvent spectrum. Another, faster, maneuver is to change the excitation wavelength by 5 or 10 nm, and if the suspect peak tracks accordingly you have probably caught the Raman peak! An example of this scenario was illustrated in Figure 4.8, which showed the emission from a dilute solution of bovine serum albumin excited at 270 nm, 280 nm, and 290 nm. One can distinguish a clear peak near 310 nm in the case of 280 nm excitation. Since tyrosine emits around here, one might imagine that the protein is exhibiting tyrosine as well as tryptophan emission. Upon 270 nm excitation, though, one sees that the "shoulder" has moved to shorter wavelengths. Excitation at 290 nm moves the suspect peak to longer wavelengths. The emission due to the protein, however, remains the same. Clearly this suspect peak was the water Raman peak and it was a coincidence (contrived by me!) that its intensity was comparable to the protein fluorescence. Raman peaks can be particularly nefarious when one is measuring polarization as a function of sample concentration. For example, consider the case of a protein labeled with fluorescein, with the goal to monitor protein dissociation, such as dimer to monomer. One may follow the polarization/anisotropy of the protein as a function of protein concentration. A convenient way to carry out this experiment is to start at as high a protein concentration as practical, determine the polarization/anisotropy, then dilute the sample twofold by removing half the sample volume and adding back the same amount of buffer. Ideally, as the protein is diluted, the polarization/anisotropy will decrease in response to oligomer dissociation. After a number of iterations of this procedure, it is not uncommon for the experimentalist to notice that the polarization/anisotropy begins to rise. In fact, this increase in polarization/anisotropy is a clear indication that the Raman peak is becoming significant. Since the Raman peak is a scattering phenomenon, it is highly polarized and, hence, when its contribution to the total signal level becomes larger, the polarization/anisotropy of the signal will increase. This effect is shown in **Figure A.1**, modified from Jameson et al. (1978) (*Rev. Sci. Instrum.* 49: 510). Of course, the concentrations at which the water Raman become significant will depend upon the absorption and emission properties of the fluorophore. The larger the fluorophore's extinction coefficient and the higher its quantum yield, the more one can dilute the sample before the Raman peak becomes evident.

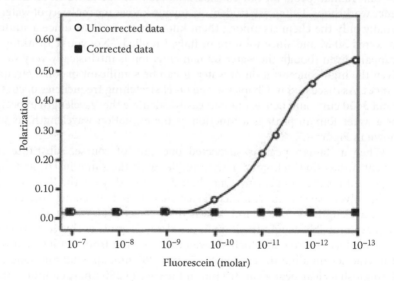

FIGURE A.1 A plot showing the measured polarization of a solution of fluorescein as a function of concentration, as well as the polarization corrected for Raman scatter.

In the example in **Figure A.1**, a solution of fluorescein in buffer was excited at 485 nm and wavelengths above ~500 nm were observed through a longpass filter. The polarization was measured as the fluorescein was diluted—the photon count rate was kept approximately constant by removing neutral density filters from the excitation pathway as the sample was diluted (the instrument utilized was actually the first photon-counting polarization instrument). As can be seen from the figure, the polarization started out near 0.02 but began to rise at fluorescein concentrations just below nanomolar levels. By 10^{-13} molar, the polarization value had surpassed 0.5. The reader will recall that 0.5 is the limit for polarization in solution and hence such high values indicate that something is amiss! At 485 nm excitation, an aqueous solution will give a Raman peak near 581 nm, which is passed by the longpass filter utilized, and since the Raman peak is highly polarized as the emission from the fluorescein decreases, the Raman contribution becomes more significant and the polarization of the solution increases. By estimating the relative contribution of fluorescein and Raman to the observed signal, and by assigning anisotropies to these contributions, namely ~0.014 for fluorescein and 0.44 for the combination of the Raman and background of the buffer, and utilizing the additivity of anisotropy rule (see Chapter 5), one is able to correct the signal for the background contribution and hence extend the measured polarization of the fluorescein down to 10^{-13} molar! Note that at some point an unpolarized contribution from unspecified fluorescent contaminants in the solvent always appears, and in the example given acts to lower the apparent polarization due to the Raman.

Inner Filter Effects

As stated earlier, one should avoid carrying out fluorescence measurements on samples with optical densities (at the exciting or emission wavelength) greater than ~0.05. Of course, this "rule of thumb" is not written in stone and polarization/anisotropy and lifetime measurements are less affected by inner filter effects than intensity measurements. The majority of fluorescence experiments involve intensity measurements, though, and it is truly amazing how often a beginner will start working with samples with high, for example, >0.5, optical densities. The consequences for certain types of experiments, especially excitation spectra, can be alarming. Titrations will often be nonlinear and in extreme cases (optical densities >2 and poor quantum yield) one may only be detecting parasitic light scattering into the detector path, since the desired exciting wavelengths have all been absorbed at the front portion of the cuvette. It is *always* good practice to take an absorption spectrum of the sample under investigation to be sure of the optical density at the exciting wavelength. If you simply *must* use a sample with a higher than desirable optical density, for example, because of its very low quantum yield, then consider using a short pathlength cuvette. **Figure A.2** illustrates the inner filter effects on the emission spectrum of fluorescein at a low (0.05; red) and high (2.0; blue) optical density at the excitation wavelength, in a 1 cm pathlength cuvette. As

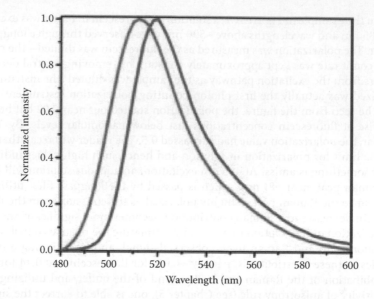

FIGURE A.2 Illustration of the inner filter effects on the emission spectrum of fluorescein at a low (0.05; pink) and high (2.0; blue) OD at the excitation wavelength, in a 1 cm pathlength cuvette. (The author would like to thank Carissa Vetromile for this figure.)

is evident, in the high optical density solution the emission spectrum is distorted due to reabsorption of the emission as it passes through the solution on the path to the detector.

Wood's Anomaly

As mentioned in Chapter 3, monochromators all have some polarization bias, that is, at any given wavelength, one plane of polarization will be transmitted more effectively than the other. Usually, this bias changes gradually with wavelength, but at some specific wavelengths, there can be a very dramatic loss of intensity in one polarization component, namely, the component oriented perpendicular to the lines of the diffraction grating. As shown in Chapter 3, this intensity loss, known as "Wood's Anomaly," can result in bumps or shoulders in emission (or excitation) spectra that are sometimes mistaken for real spectroscopic parameters. Since the position of the Wood's Anomaly in the emission spectrum will not change upon a change in excitation wavelength, one can understand why it could be construed as an intrinsic part of the emission. If the operator of the fluorimeter is not completely familiar with the instrument being utilized, they may not know the position of the Wood's Anomalies. If the fluorimeter is equipped with polarizers, one can simply place a polarizer in the emission path and observe what happens to the emission spectrum as spectra are scanned with the polarizer oriented first in one direction and then the other direction. This aspect was illustrated in Figure 3.17 and also Figure 4.9. If the dip or shoulder is a Wood's Anomaly, one will note that it disappears for one orientation of the polarizer, and that it is more pronounced for the other polarizer orientation. Since instrument manufacturers do not follow a common guideline regarding the orientation of the monochromators, that is, most are oriented so that the lines of the grating are parallel to the vertical laboratory axis, but some are oriented so that the lines of the grating are perpendicular to the vertical laboratory axis, one should try both polarizer orientations to be sure. *Anecdote alert! I remember clearly my introduction to Wood's Anomaly. I was testing a new red sensitive PMT (the Hamamatsu 928) that we had just received in Weber's lab and I was looking at a rhodamine emission, which had a maximum near 630 nm. I noticed a dip in the emission near the maximum wavelength that I had never seen before—since our previous PMT did not have much sensitivity above 600 nm. I noted that the dip disappeared when viewed through a parallel polarizer and was more pronounced when viewed through a perpendicular polarizer. I certainly had no idea what was going on and then Gregorio Weber came into the room and asked what I was up to. I explained my observations and then he turned to leave—and as the door was closing behind him, he said, "That is the well-known Wood's Anomaly." As I said earlier in this book, we did not have Google then—we had Weber!*

Photobleaching

Photobleaching of samples is most commonly associated with microscopy measurements—sometimes this phenomenon is useful, such as in FRAP (Chapter 9), but usually not. For *in vitro* studies, one rarely wants to see photobleaching. Often, however, the novice pumps too much exciting light into the sample and photobleaching of the fluorophore occurs. This problem can easily arise when one works with a sample that does not diffuse rapidly out of the excitation beam, for example, large proteins or protein aggregates. Tryptophan, for example, is prone to photobleach under strong UV illumination. I advise anyone carrying out fluorescence studies on an unfamiliar sample to be alert to the possibility of photobleaching. One should set up the experiment, for example, an emission spectrum, with the slitwidth and other instrument parameters of choice, and then observe the signal intensity, preferably at the emission maximum, for a minute or so to see if the intensity decreases with time. If a decrease is observed, one may close the excitation shutter for 10 s or so and then open it to see if the intensity has returned to its original value, that is, the value before the bleaching occurred. One has then witnessed fluorescence recovery after photobleaching! To prevent photobleaching in the system, one must reduce the intensity of the exciting light. The way in which you accomplish this goal will vary depending on your particular instrumentation. In some cases, you may be able to lower the lamp current; in others, you can decrease the excitation slitwidth. In still others, you can place a neutral density filter in the excitation pathway (my favorite method). The point is to reduce the excitation light and then once again observe the signal over time to see if photobleaching still occurs. If signal is a problem, one can usually increase the emission slits or the emission PMT voltage (in analog systems).

Mistaking Second-Order Scatter for Fluorescence, Both in Emission and Excitation Spectra

I have noticed that in recent years, more and more novices tend to take a fluorescence spectrum over a very wide range of emission wavelengths. This approach may be due to the default settings on some spectrofluorimeters—I don't really know. For example, I have seen people excite a sample at 280 nm but then run the emission spectrum from 250 nm up to 750 nm. They often seem to recognize that the 280 nm peak corresponds to the exciting light (the Rayleigh scatter), but sometimes fail to realize that the peak at 560 nm is just the second-order scattering of the exciting light. Of course, if they were to scan to even longer emission wavelengths they might even observe a third-order peak near 840 nm! Second-order peaks are illustrated in **Figure A.3**. In this example, a dilute tryptophan solution was excited at 270 nm and the sample was scanned from 250 to 800 nm. One notes that the Rayleigh peak, the Raman peak, and the tryptophan fluorescence, at 270 nm, 297 nm, and near 350 nm, are all repeated at higher wavelengths, although at lower intensities.

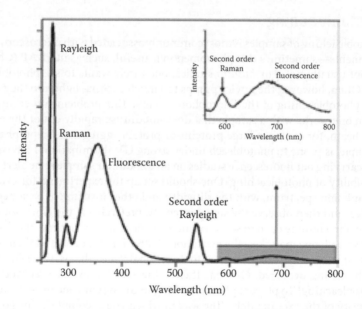

FIGURE A.3 Illustration of second-order peaks in a broad emission scan. Excitation of a dilute aqueous solution of tryptophan at 270 nm gives rise to a Rayleigh peak, a Raman peak, tryptophan emission and then, above 500 nm, a repeat of these peaks as second-order effects.

One also notes that the center of the tryptophan second order is a bit less than twice the main emission, indicating that the second-order outputs can be slightly nonlinear (at least in the two instruments I tested). Also, note that this effect is not just due to PMT correction factors, since the second-order emission peak is *not* near 700 nm, it is actually near 350 nm.

Photon Counting and Pulse Pileup

As mentioned in Chapter 3, the photon counting method is the most sensitive way to detect and quantify light. Photon counting is used in the time-domain lifetime technique of TCSPC. It is also used in many steady-state spectrofluorimeters. Although photon counting offers superb sensitivity, it presents some problems when the signal is large. Specifically, if the count rate is too high, the phenomenon of pulse pileup can occur. As shown in Figure 3.24, photons are converted into electronic pulses of a finite width, and the counting electronics are essentially blind to any photons that arrive during the time period corresponding to an active pulse width. Consequently, if one were to monitor the count rate as a signal got larger and larger, one would see that the expected linearity between incoming photons and counts begins to break down. Different instruments will exhibit pulse pileup effects at different count rates since the effect depends upon the details of the electronics, for example, pulse width. Hence, one should ask the manufacturer, or better yet carry out some simple tests for oneself. In addition to compromising spectra, pulse pileup can give rise to erroneous polarization/anisotropy values. Typically, the pileup tends to lower polarization/anisotropy values since the pileup will occur first in the reading which gives more counts, which is usually the parallel emission channel (in the case of positive values).

Errors in Determining Relative Fluorescence Intensities and Quantum Yields

A very common mistake involving simple fluorescence intensity measurements, for example, during quenching or protein denaturation experiments, is to monitor the emission at a single wavelength and to report the change in intensity at that wavelength. Of course, if there is any shift in the emission maximum during the experimental intervention, it is obvious that the intensities will be skewed and will not represent simply the change in quantum yield. But there is a more insidious error lying in wait. Namely, if the experimental manipulation alters the polarization/anisotropy of the emission, which can easily happen if the lifetime or fluorophore environment is altered, than the normal right angle observation geometry will introduce a polarization bias into the intensities. Thus, bias results since there will be two perpendicularly polarized components to the emission, but only one can be detected. One way to avoid this artifact is to excite the sample with vertically polarized light and then record the emission intensities viewed through both vertically and horizontally oriented polarizers; then calculate the sum of the parallel intensity plus twice the perpendicular intensities (i.e., $I_{\parallel} + 2\,I_{\perp}$). This sum represents the *total* emission intensity and removes any polarization bias. Alternatively, one can excite the sample using one of the "magic angle" conditions mentioned in Chapter 6. For example, one may excite the sample with parallel polarized light and view the emission through an emission polarizer set at 55°—this procedure must, of course, be carried out for the sample in the presence and absence of the perturbation. Also, as mentioned in Chapter 7, one should not use simple OD (optical density) values when calculating relative quantum yields. Rather one must use the fraction of photons absorbed by the sample and reference solutions, that is,

$$QY_{sample} = QY_{standard}\,(I_{T\,sample}/I_{T\,standard})(1 - 10^{-ODsample}/1 - 10^{-ODstandard})$$
$$\times\,(n_{sample}/n_{standard})$$

Incorrect Use of Polarization Data

I was greatly surprised to note that some researchers deal with polarization data as if it were completely equivalent to anisotropy data. Of course, the basic information content of these two functions is essentially equivalent, but the use of one or the other term in certain types of data manipulations requires attention. The classic error is to treat polarization data as if it were linearly additive the same as anisotropy data. As discussed in Chapter 5, the anisotropy of a mixture of fluorophores will correspond to the sum of the anisotropies of the different emitting species in the solution, weighted according to their fractional contribution to the total intensity. Another error sometimes made is the subtraction of the polarization (or anisotropy) of the background directly from the polarization (or anisotropy) of the sample. As discussed in Chapter 5, to correct the sample polarization for background, one must subtract the value of the background for the different positions of the polarizers from the values observed for the sample with those same polarizer settings, as was indicated in Equation 5.20.

Unwarranted Use of 2/3 for Kappa Square

A man out walking one dark night comes upon another man circling a lamppost searching the ground beneath it. He asks the man what he is looking for and the reply is, "I lost my keys." The first man, being a good Samaritan, wants to help look and casually says, "So you lost them somewhere around here?" To this, the other man replies, "No, I lost them down the street, but the light is better here." This anecdote reminds me of the fact that the use of 2/3 for κ^2 in FRET calculations is almost universal among the casual FRET practitioners: 2/3 may not be the correct value but it's the easiest one to use! Of course, this circumstance owes itself to the difficulty in accurately determining κ^2. In some cases, of course, the 2/3 values is reasonable—particularly when the donor and acceptor are small molecules, that is, are not fluorescent proteins like GFP, and when they are free to move significantly during the excited state lifetime. These issues were discussed in some detail in Chapter 9.

Effects of Turbidity and Scattering on Polarization

As mentioned earlier, sample turbidity will actually decrease the observed polarization due to multiple scattering of the emission. Typically, novices encounter two problems when measuring polarization/anisotropy on turbid (even slightly turbid) samples. These two problems have opposite effects, namely (1) an increase in apparent polarization/anisotropy due to parasitic light reaching the detector, and (2) a decrease in the apparent polarization/anisotropy due to depolarization of the excitation and emission light via scattering. To check for the first problem, one needs only to use a proper background solution, that is, the same preparation minus the fluorophores, in the same type of cuvette used to measure the sample and determine if any signal is detected. If the proper background solution is used, any parasitic light reaching the detector will be evident. One can then try using an appropriate interference filter in the excitation pathway to remove the parasitic light. After checking for this problem, one can then assess the issue of depolarization of the emission via scattering processes. The easiest way to accomplish this goal is probably to carry out the measurement in different pathlength cuvettes, for example, a 3 mm × 3 mm cuvette compared to a 10 mm × 10 mm cuvette. If the polarization determined using a shorter pathlength cuvette is higher than that found with a longer pathlength cuvette, then one has identified a problem. The effect of scattering of the fluorescence on anisotropy was demonstrated in Figure 5.27.

Incorrect Filter Choice

I refer here to filter choices for *in vitro* measurements, not microscopy. Sometimes, one has a sample with relatively weak fluorescence compared to the exciting light, and the filter being utilized may not totally block the Rayleigh scatter. One must remember that longpass filters are not perfect optical devices. Even if the transmission cut-on looks very sharp, there can be a small fraction of light passed at wavelengths below the apparent cut-on wavelength. For this reason, it is best to check your filter choice by setting up for the measurement and then moving the filter in the emission side over to the excitation path to verify that no exciting light gets through. If the filter is good, the observed signal, with the emission filter in the excitation path, should be around the dark count level.

Index

Index

Index

T - #0022 - 101024 - C314 - 234/156/17 [19] - CB - 9781439806043 - Gloss Lamination